Noncommutative Maslov Index and Eta-Forms

Memoirs
of the
American Mathematical Society

Number 887

Noncommutative Maslov Index and Eta-Forms

Charlotte Wahl

2000 *Mathematics Subject Classification.*
Primary 19K56; Secondary 53D12, 58J28, 46L87.

Library of Congress Cataloging-in-Publication Data
Wahl, Charlotte, 1972–
　Noncommutative Maslov index and eta-forms / Charlotte Wahl.
　　p. cm. — (Memoirs of the American Mathematical Society, ISSN 0065-9266 ; no. 887)
　"September 2007, volume 189, number 887 (end of volume)."
　Includes bibliographical references.
　ISBN 978-0-8218-3997-3 (alk. paper)
　1. Index theory (Mathematics). 2. Maslov index. 3. K-theory. I. Title.
QA614.92.W34　2007
514′.74—dc22
　　2007060803

Memoirs of the American Mathematical Society

This journal is devoted entirely to research in pure and applied mathematics.

Subscription information. The 2007 subscription begins with volume 185 and consists of six mailings, each containing one or more numbers. Subscription prices for 2007 are US$649 list, US$519 institutional member. A late charge of 10% of the subscription price will be imposed on orders received from nonmembers after January 1 of the subscription year. Subscribers outside the United States and India must pay a postage surcharge of US$38; subscribers in India must pay a postage surcharge of US$43. Expedited delivery to destinations in North America US$53; elsewhere US$130. Each number may be ordered separately; *please specify number* when ordering an individual number. For prices and titles of recently released numbers, see the New Publications sections of the *Notices of the American Mathematical Society*.

Back number information. For back issues see the *AMS Catalog of Publications*.

Subscriptions and orders should be addressed to the American Mathematical Society, P. O. Box 845904, Boston, MA 02284-5904, USA. *All orders must be accompanied by payment.* Other correspondence should be addressed to 201 Charles Street, Providence, RI 02904-2294, USA.

Copying and reprinting. Individual readers of this publication, and nonprofit libraries acting for them, are permitted to make fair use of the material, such as to copy a chapter for use in teaching or research. Permission is granted to quote brief passages from this publication in reviews, provided the customary acknowledgment of the source is given.

Republication, systematic copying, or multiple reproduction of any material in this publication is permitted only under license from the American Mathematical Society. Requests for such permission should be addressed to the Acquisitions Department, American Mathematical Society, 201 Charles Street, Providence, Rhode Island 02904-2294, USA. Requests can also be made by e-mail to `reprint-permission@ams.org`.

Memoirs of the American Mathematical Society is published bimonthly (each volume consisting usually of more than one number) by the American Mathematical Society at 201 Charles Street, Providence, RI 02904-2294, USA. Periodicals postage paid at Providence, RI. Postmaster: Send address changes to Memoirs, American Mathematical Society, 201 Charles Street, Providence, RI 02904-2294, USA.

© 2007 by the American Mathematical Society. All rights reserved.
Copyright of this publication reverts to the public domain 28 years
after publication. Contact the AMS for copyright status.
This publication is indexed in *Science Citation Index*®, *SciSearch*®, *Research Alert*®,
CompuMath Citation Index®, *Current Contents*®/*Physical, Chemical & Earth Sciences*.
Printed in the United States of America.

∞ The paper used in this book is acid-free and falls within the guidelines
established to ensure permanence and durability.
Visit the AMS home page at `http://www.ams.org/`

10 9 8 7 6 5 4 3 2 1　　12 11 10 09 08 07

Contents

Introduction	1
Summary	4
Notation and conventions	5
Chapter 1. Preliminaries	7
1.1. The geometric situation	7
1.2. The family index theorem	9
1.3. The algebra of differential forms	10
1.4. Lagrangian projections	16
Chapter 2. The Fredholm Operator and Its Index	21
2.1. The operator D on M	21
2.2. The operator D_I on $[0,1]$	23
2.3. The operator D_Z on the cylinder	27
2.4. The index of D^+	28
2.5. A perturbation with closed range	31
Chapter 3. Heat Semigroups and Kernels	33
3.1. Complex heat kernels	33
3.2. The heat semigroup on closed manifolds	38
3.3. The heat semigroup on $[0,1]$	40
3.4. The heat semigroup on the cylinder	44
3.5. The heat semigroup on M	49
Chapter 4. Superconnections and the Index Theorem	61
4.1. The superconnection A_t^I associated to D_I	61
4.2. The superconnection associated to D_Z	68
4.3. The superconnection $A(\rho)_t$ associated to $D(\rho)$	69
4.4. The index theorem and its proof	72
4.5. A gluing formula for η-forms on S^1	86
Chapter 5. Definitions and Techniques	91
5.1. Hilbert C^*-modules	91
5.2. Operators on spaces of vector valued functions	98
5.3. Projective systems and function spaces	105
5.4. Holomorphic semigroups	110
Bibliography	117

Abstract

We define and prove a noncommutative generalization of a formula relating the Maslov index of a triple of Lagrangian subspaces of a symplectic vector space to eta-invariants associated to a pair of Lagrangian subspaces. The noncommutative Maslov index, defined for modules over a C^*-algebra $\mathcal{A}$, is an element in $K_0(\mathcal{A})$. The generalized formula calculates its Chern character in the de Rham homology of certain dense subalgebras of $\mathcal{A}$. The proof is a noncommutative Atiyah-Patodi-Singer index theorem for a particular Dirac operator twisted by an $\mathcal{A}$-vector bundle. We develop an analytic framework for this type of index problem.

Received by the editor 06/08/2005.
2000 *Mathematics Subject Classification.* 19K56 (53D12, 58J28, 46L87).
Key words and phrases. Dirac operator, Maslov index, eta-invariant, higher index theory, K-theory.

Introduction

The purpose of this paper is twofold. We establish a noncommutative generalization of a formula by Cappell, Lee and Miller [**CLM**] relating the Maslov index of a triple of Lagrangian subspaces of a symplectic vector space to certain η-invariants which can be associated to a pair of Lagrangian subspaces. The formula was generalized to families of Lagrangian subspaces by Bunke and Koch [**BK**]. Our Maslov index will be defined for C^*-modules and will be an element in the K-theory of the C^*-algebra, whereas the η-invariants are replaced by noncommutative differential forms. The proof of our formula is a noncommutative index theorem for a particular Dirac operator twisted by a C^*-vector bundle on a two-dimensional manifold with boundary and cylindric ends. The index theorem calculates the Chern character of index of the Dirac operator in the de Rham homology of certain dense subalgebras of the C^*-algebra.

The second aim is to provide a precise analytic framework for the index theory of Dirac operators over a certain type of dense subalgebras of C^*-algebras based on heat kernel methods and the Quillen superconnection formalism. In particular these dense subalgebras are assumed to be projective limits of Banach algebras, so that we mostly deal with Banach space valued functions. Our results apply to higher index theory for invariant Dirac operators on covering spaces. Proofs in higher index theory based on the superconnection formalism have been given before [**Lo1**] [**Lo2**] [**LP1**] [**LP2**] [**Wu**]. Some analytical methods relevant for the general situation have been developed by Lott in [**Lo3**].

The Maslov index $\tau(L_0, L_1, L_2)$ of a triple of Lagrangian subspaces (L_0, L_1, L_2) of $\mathbb{R}^{2n}$ endowed with the standard symplectic form ω is defined as the signature of the quadratic form q on $L_0 \cap (L_1 + L_2)$ [**LV**] given by

$$q(x_1 + x_2) := \omega(x_2, x_1), \ x_i \in L_i \ .$$

Its geometric significance comes from a gluing formula for signatures of manifolds with boundary [**Wa**]: Let M^{4n} be an oriented manifold with boundary and let N be a hypersurface in M with boundary. Cutting along N produces two (topological) manifolds M_1, M_2 with boundary. The images of $H^{2n-1}(N)$ and $H^{2n-1}(M_i)$, $i = 1, 2$, in $H^{2n-1}(\partial N)$ are Lagrangian subspaces with respect to the symplectic form induced by the cup product. Up to sign the difference of the signatures $\sigma(M) - \sigma(M_1) - \sigma(M_2)$ equals the Maslov index of these subspaces.

The η-invariant associated to a pair (L_0, L_1) of Lagrangian subspaces is defined as the η-invariant of the operator $D_I = I_0 \frac{d}{dx}$ on $L^2([0, 1], \mathbb{R}^{2n})$ with boundary conditions $f(0) \in L_0$, $f(1) \in L_1$. Here I_0 is the skewsymmetric matrix representing the symplectic form ω with respect to the standard scalar product on $\mathbb{R}^{2n}$. Then

$$\eta(L_0, L_1) = \frac{1}{\sqrt{\pi}} \int_0^\infty t^{-\frac{1}{2}} \operatorname{Tr} D_I e^{-tD_I^2} dt \ .$$

It is a regularization of the difference between the number of positive and negative eigenvalues of D_I. The η-invariant $\eta(L_0, L_1)$ occurs as a correction term in a gluing theorem for η-invariants [**Bu**]. The formula

$$\tau(L_0, L_1, L_2) = \eta(L_0, L_1) + \eta(L_1, L_2) + \eta(L_2, L_0) \ ,$$

which will be generalized to a noncommutative context, can be interpreted as reflecting the connection between the gluing problems for signatures and η-invariants via the Atiyah-Patodi-Singer index theorem.

The noncommutative Maslov index and the η-forms should play a similar role in gluing problems for higher signatures for manifolds with boundaries [**LP3**], [**LLP**], [**LLK**], higher η-invariants and ρ-invariants [**Lo1**]. In §4.5 we explain how our formula is related to a gluing formula for noncommutative η-forms for Dirac operators on the circle.

The proof by Cappell, Lee and Miller relies on the axiomatic properties of the Maslov index, which are not expected to hold for a noncommutative generalization. Bunke and Koch [**BK**] formulated an index problem whose solution, an Atiyah-Patodi-Singer type index theorem, yields the above formula. They interpreted the Maslov index as the index of a Dirac operator on a two-dimensional manifold with boundary and six cylindric ends isometric to $\mathbb{R}^+ \times [0,1]$. The Lagrangian subspaces enter in the definition of the boundary conditions. The purpose of their approach was to prove a formula for families of Lagrangian subspaces.

Our proof is a generalization of the approach of Bunke and Koch. Instead of a family of Dirac operators we consider a Dirac operator twisted by a C^*-vector bundle [**MF**].

Before stating the result we give a short introduction in the type of noncommutative index theorem considered in this paper:

Family index theorems describe Fredholm operators depending continuously on a parameter from some compact space. The index of a family of operators is an element in K-theory of the base space. If the kernel and the cokernel are vector bundles, then the index is the difference of the classes of these bundles.

One may reformulate this situation by replacing the base space B by the C^*-algebra $C(B)$ and the family of operators by an operator on a $C(B)$-module. The index is then in the C^*-algebraic K-theory $K_0(C(B))$, which is naturally isomorphic to $K^0(B)$.

Miščenko and Fomenko [**MF**] formalized and generalized this point of view by defining Fredholm operators on Hilbert C^*-modules for general C^*-algebras. The index of such a Fredholm operator is an element in the K-theory of the C^*-algebra. They also elaborated a theory of pseudodifferential operators over C^*-algebras. Important examples for geometric applications are Dirac operators associated to C^*-vector bundles.

In order to formulate a noncommutative analogue of the family index theorem, which calculates the Chern character of the index of a family of Dirac operators in the cohomology of the base space, now assumed to be a manifold, analogues of differential forms, de Rham cohomology and the Chern character are needed. Karoubi [**Ka**] introduced a complex of differential forms $(\hat\Omega_*\mathcal{A}, \mathrm{d})$ associated to a Fréchet algebra $\mathcal{A}$, its de Rham homology $H^{dR}_*(\mathcal{A})$ and a Chern character $\mathrm{ch}: K_0(\mathcal{A}) \to H^{dR}_*(\mathcal{A})$, which is defined if in addition $\mathcal{A}$ is a local Banach algebra. Unfortunately the de Rham homology of a C^*-algebra does not behave well, in particular the de Rham homology of a commutative unital C^*-algebra is in general not the cohomology of the corresponding compact space. By considering the de Rham homology of the algebra $C^\infty(B)$ instead of $C(B)$ one recovers the Chern character from differential geometry. This can be interpreted as reflecting the fact that a differentiable structure on B is needed for the definition of de Rham cohomology.

In order to get a reasonable Chern character in the general situation one chooses a dense subalgebra $\mathcal{A}_\infty$ holomorphically closed under the functional calculus in $\mathcal{A}$. Then $K_0(\mathcal{A}_\infty)$ is canonically isomorphic to $K_0(\mathcal{A})$, and the Chern character yields a homomorphism

$$\mathrm{ch}: K_0(\mathcal{A}) \to H^{dR}_*(\mathcal{A}_\infty) \ .$$

In this setting homological index theorems for Dirac operators associated to C^*-vector bundles can be formulated and – at least formally – a noncommutative version of Bismut's family index theorem makes sense [**BGV**]. Note that formally the noncommutative situation is less complex than the family case since the Riemannian metric is fixed.

For the generalization of the heat kernel theory additional conditions have to be imposed on the algebra $\mathcal{A}_\infty$: It should be the limit of a projective system $\{\mathcal{A}_i\}_{i\in\mathbb{N}_0}$ of involutive Banach algebras with $\mathcal{A}_0 = \mathcal{A}$, such that there are dense embeddings $\mathcal{A}_{i+1} \hookrightarrow \mathcal{A}_i$ and such that each $\mathcal{A}_i$ is closed with respect to the holomorphic functional calculus in $\mathcal{A}$. The motivating example is $\mathcal{A}_i := C^i(B)$ for a closed manifold B. More generally, if δ is an involutive closed derivation on $\mathcal{A}$ with $\cap_i \mathrm{dom}\, \delta^i$ dense in $\mathcal{A}$, then the projective system given by $\mathcal{A}_i := \mathrm{dom}\, \delta^i$ with norm

$$\|a\|_i := \sum_{j=0}^{i} \|\delta^j(a)\|$$

fulfills the conditions. For a discrete finitely generated group G the unbounded operator D on $l^2(G)$ defined by $D1_g = l(g)1_g$, where l is a word length function on G, induces a closable derivation $\delta(f) = [D, f]$ on $B(l^2(G))$. Furthermore $\mathbb{C}G \subset \mathcal{A}_i := \mathrm{dom}\, \overline{\delta}^i \cap C^*_r(G)$. The projective limit $\mathcal{A}_\infty \subset C^*_r(G)$ is closely related to the algebra employed by Connes and Moscovici in their proof of the Novikov conjecture for Gromov hyperbolic groups via higher index theory [**CM**]. Using this setting our results can be applied to higher index theory. A proof of a higher Atiyah-Singer index theorem based on heat kernel methods was given by Lott [**Lo2**] and a higher Atiyah-Patodi-Singer index theorem was proved by Leichtnam and Piazza [**LP1**] [**LP2**]. Lott introduced higher η-invariants and ρ-invariants [**Lo1**]. A motivation for higher index theory is the study of higher signatures and the Novikov conjecture. Index theory for Dirac operators over C^*-algebras in general has been applied to the study of manifolds with positive scalar curvature [**PS**] [**Ros2**].

The proofs of the higher index theorems mentioned above rely on the comparison with the situation on the covering space where one can deal with operators on sections of complex vector bundles. In the general setting the main difficulty lies in the fact that the calculus of regular operators on a Hilbert C^*-module and the calculus of pseudodifferential operators over a C^*-algebra are not sufficient for the study of the heat semigroup since we have to deal with vector bundles whose fibers are projective $\mathcal{A}_\infty$-modules. By proving that the de Rham homology behaves well under the projective limit we justify the fact that one can deal with Banach spaces instead of Fréchet spaces. Then the theory of holomorphic semigroups can be used instead of the calculus of selfadjoint operators. We define appropriate operator spaces on L^2-spaces of vector valued functions, for example Hilbert-Schmidt operators and trace class operators. Duhamel's principle is used for the construction and the short time asymptotics of the heat kernels. The long time asymptotics is more intricate: here we have to get hold of the spectrum of the Dirac operator. We adapt a method developed by Lott, namely a restricted pseudodifferential operator calculus giving information about the resolvent set of the Dirac operator and the regularizing properties of the resolvents [**Lo3**].

Although the theory is developed only for a particular two-dimensional manifold and a trivial vector bundle, the relevant parts generalize to the Atiyah-Patodi-Singer index problem for Dirac operators over C^*-algebras, which will be considered elsewhere.

Now we can formulate the noncommutative version of the equation relating the Maslov index and the η-invariants.

Let $\mathbb{C}^{2n}$ be endowed with the skewhermitian form ω induced by the standard symplectic form on $\mathbb{R}^{2n}$ via the identification $\mathbb{C}^{2n} = \mathbb{R}^{2n} \otimes_\mathbb{R} \mathbb{C}$.

For a C^*-algebra $\mathcal{A}$ a Lagrangian projection on $\mathcal{A}^{2n}$ is a selfadjoint projection $P \in M_{2n}(\mathcal{A})$ fulfilling $PI_0 = (1 - I_0)P$. Two projections are called transverse if their sum is invertible.

As above one associates an $\mathcal{A}$-valued hermitian form q to every triple $(\mathcal{P}_0, \mathcal{P}_1, \mathcal{P}_2)$ of pairwise transverse Lagrangian projections. The class $\tau(\mathcal{P}_0, \mathcal{P}_1, \mathcal{P}_2) := [q] \in K_0(\mathcal{A})$ is called Maslov index of $(\mathcal{P}_0, \mathcal{P}_1, \mathcal{P}_2)$.

An η-form $\eta(\mathcal{P}_0, \mathcal{P}_1) \in \hat{\Omega}_*\mathcal{A}_\infty / \overline{[\hat{\Omega}_*\mathcal{A}_\infty, \hat{\Omega}_*\mathcal{A}_\infty]_s}$ can be associated to a pair of transverse Lagrangian projections $(\mathcal{P}_0, \mathcal{P}_1)$ with $\mathcal{P}_i \in M_{2n}(\mathcal{A}_\infty)$, $i = 0, 1$. Here $[\,,\,]_s$ denotes the supercommutator. The η-form is defined via a superconnection associated to the operator $D_I = I_0 \frac{d}{dx}$ on $L^2([0,1], \mathcal{A}^{2n})$ with boundary conditions $f(0) = \mathcal{P}_0 f(0)$ and $f(1) = \mathcal{P}_1 f(1)$.

Then our main result is:

THEOREM. *For a triple $(\mathcal{P}_0, \mathcal{P}_1, \mathcal{P}_2)$ of pairwise transverse Lagrangian projections with $\mathcal{P}_i \in M_{2n}(\mathcal{A}_\infty)$, $i = 0, 1, 2$,*

$$\operatorname{ch}\tau(\mathcal{P}_0, \mathcal{P}_1, \mathcal{P}_2) = [\eta(\mathcal{P}_0, \mathcal{P}_1) + \eta(\mathcal{P}_1, \mathcal{P}_2) + \eta(\mathcal{P}_2, \mathcal{P}_0)] \in H^{dR}_*(\mathcal{A}_\infty) \ .$$

This paper is based on the author's PhD thesis. I would like to thank my supervisor Ulrich Bunke for drawing my attention to the problem and for fruitful discussions, and Margit Rösler for the introduction into the theory of semigroups.

Summary

The paper is organized in the following way:

As mentioned above, the Maslov index $\tau(\mathcal{P}_0, \mathcal{P}_1, \mathcal{P}_2)$ is the index of a Dirac operator D^+ twisted by a C^*-vector bundle on a two-dimensional spin manifold M with six cylindric ends isometric to $[0, \infty) \times [0, 1]$.

In Chapter 1 the manifold M is described and we explain in more detail the family index theorem of Bunke and Koch [**BK**]. Then the universal differential algebra $\hat{\Omega}_*\mathcal{A}_\infty$, the de Rham homology $H_*^{dR}(\mathcal{A}_\infty)$ and the Chern character are introduced and investigated. It is shown that $H_*^{dR}(\mathcal{A}_\infty)$ is the projective limit of $H_*^{dR}(\mathcal{A}_i)$. Moreover Lagrangian projections and the Maslov index are defined.

In Chapter 2 we introduce the Dirac operator $D = D^+ \oplus D^-$ on M whose boundary conditions are defined by a triple of Lagrangian projections $(\mathcal{P}_0, \mathcal{P}_1, \mathcal{P}_2)$. Furthermore the operator D_I is defined and its properties on the Hilbert $\mathcal{A}$-module $L^2([0,1], \mathcal{A}^{2n})$ are studied. Then we show that D^+ is Fredholm between appropriate Hilbert C^*-modules and that its index equals $\tau(\mathcal{P}_0, \mathcal{P}_1, \mathcal{P}_2)$. We define a compact perturbation $D(\rho)$ of D such that $D(\rho)^+$ is surjective. Then the index of D^+ can be expressed in terms of the kernel of $D(\rho)$.

Chapter 3 is devoted to heat semigroups and their integral kernels, in particular those associated to D_I and $D(\rho)$.

In Chapter 4 we introduce superconnections in order to define the η-form. Now we can formulate the index theorem. The remainder of the chapter is devoted to its proof. We introduce the rescaled superconnection $A(\rho)_t$ associated to $D(\rho)$ and study the family of operators $e^{-A(\rho)_t^2}$. We show that it is a family of integral operators with smooth integral kernel and obtain estimates for the integral kernel for small t. Once the heat kernel theory is established, the proof of the index theorem itself is fairly standard. We follow the proof in [**BK**], which is modelled on Melrose's b-calculus [**Me**], and compare the limit of a generalized supertrace of $e^{-A(\rho)_t^2}$ for $t \to \infty$, which is the Chern character of the index of D^+, with its limit for $t \to 0$. Since the differences between the calculus for Dirac operators associated to complex vector bundles and the one for Dirac operators associated to a projective system of vector bundles as developed here are subtle, the proof is given in detail.

In Chapter 5 the functional analytic framework is developed. The function spaces we deal with are introduced, as for example the L^2-spaces of vector valued functions, and operators on them are studied. In particular we define and study appropriate notions of adjointable operators, Hilbert-Schmidt and trace class operators. Furthermore we recall the properties of Fredholm operators and regular operators on Hilbert C^*-modules, and collect some facts from holomorphic semigroup theory. The reader is advised to go through the definitions of this chapter first in order to get acquainted with the functional analytic setting.

Notation and conventions

If not specified vector spaces and algebras are complex, manifolds are smooth.

We often deal with $\mathbb{Z}/2$-graded spaces. Then $[\,,\,]_s$ denotes the supercommutator and tr_s the supertrace. In a graded context the tensor products are graded. For an ungraded vector space V, we denote by V^+ resp. V^- the same space endowed with a grading: all elements are homogeneous of positive resp. negative degree.

Tensor products denoted by $\otimes$ are completed. The way of completion is indicated by a suffix in all but the two most common cases: In the case of Hilbert C^*-modules $\otimes$ means the Hilbert C^*-module tensor product, and if one of the spaces is a nuclear locally convex space, then $\otimes$ means $\otimes_\pi$ or $\otimes_\varepsilon$. The algebraic tensor product is denoted by $\odot$.

By a differentiable function on an open subset of $[0,1]^n$ we understand a function that can be extended to a differentiable function on an open subset of $\mathbb{R}^n$.

This induces the notion of a differentiable function on a manifold with corners, in our case $M \times M$.

If S, X are sets with $S \subset X$, then the characteristic function of S is denoted by $1_S : X \to \{0,1\}$. If X is a metric space, $y \in X$ and $S \subset X$ then $d(y, S) := \inf_{x \in S} d(x, y)$. For $S_1, S_2 \subset X$ we set $d(S_1, S_2) := \inf_{x \in S_1} d(x, S_2)$.

If E_i, $i = 1, 2$, is a vector bundle on a space X_i, $i = 1, 2$, and $p_i : X_1 \times X_2 \to X_i$ is the projection, then $E_1 \boxtimes E_2 = p_1^* E_1 \otimes p_2^* E_2$ on $X_1 \times X_2$.

We use the notions from [**BGV**] in the context of spin geometry. In addition let a Dirac bundle be a selfadjoint Clifford module endowed with a Clifford connection with respect to which the metric is parallel. Our sign conventions differ from those in [**BGV**] since we deal with right modules over the algebra of differential forms.

The value of the constant C used in estimates may vary during a series of estimates without an explicit remark.

CHAPTER 1

Preliminaries

1.1. The geometric situation

In this section the two-dimensional spin manifold M with boundary and cylindric ends, a Dirac bundle E on it and the associated Dirac operator will be introduced. The open covering $\mathcal{U}(r,b)$ of M constructed in this section will be used for cutting and pasting arguments later on. Furthermore we will fix a flat open set $F \subset M$ containing the boundary and the cylindric ends and trivializations of $TM|_F$ and of $E|_F$, which will be used in the definition of the boundary conditions for the Dirac operator in §2.1.1.

We begin by defining the manifold M.

For $k \in \mathbb{Z}/6$ let Z_k be a copy of $\mathbb{R} \times [0,1]$. With the euclidian metric and the standard orientation Z_k is an oriented Riemannian manifold with boundary. Let (x_1^k, x_2^k) be the euclidian coordinates of Z_k.

For $r \geq -\frac{1}{2}$, $b \leq \frac{1}{3}$ and $k \in \mathbb{Z}/6$ let

$$F_k(r,b) := \{(x_1^k, x_2^k) \in\,]r, \infty[\times[0,1]\, \cup\,]-1, r] \times ([0,b[\,\cup\,]1-b, 1])\} \subset Z_k\, .$$

We define

$$F(r,b) := \Big(\bigcup_{k \in \mathbb{Z}/6} F_k(r,b)\Big)/\sim$$

with $(x_1^k, x_2^k) \sim (-\frac{3}{2} - x_1^{k-1}, 1 - x_2^{k-1})$ for $(x_1^k, x_2^k) \in\,]-1, -\frac{1}{2}[\times[0,b[$ and $k \in \mathbb{Z}/6$.

Then $F(r,b)$ inherits the structure of an oriented Riemannian manifold from the sets $F_k(r,b)$.

The set $F(-\frac{1}{2}, \frac{1}{3})\backslash\overline{F(-\frac{1}{3}, \frac{1}{4})}$ is diffeomorphic to the open ring $B_1(0)^\circ \backslash B_{1/2}(0) \subset \mathbb{R}^2$ via an oriented diffeomorphism ϕ. We define the manifold with boundary

$$M := F(-\tfrac{1}{2}, \tfrac{1}{3}) \cup_\phi B_1(0)^\circ\, .$$

For $r > -\frac{1}{2}$, $b \leq \frac{1}{3}$ we identify $F(r,b)$ and $F_k(r,b)$ with the corresponding subsets in M. The sets $F_k(r,b)$ are coordinate patches of M with the coordinates (x_1^k, x_2^k) from above.

Extend the orientation and metric from $F := F(0, \frac{1}{4})$ to the whole of M and endow TM with the Levi-Civita connection. In the following we identify TM and T^*M.

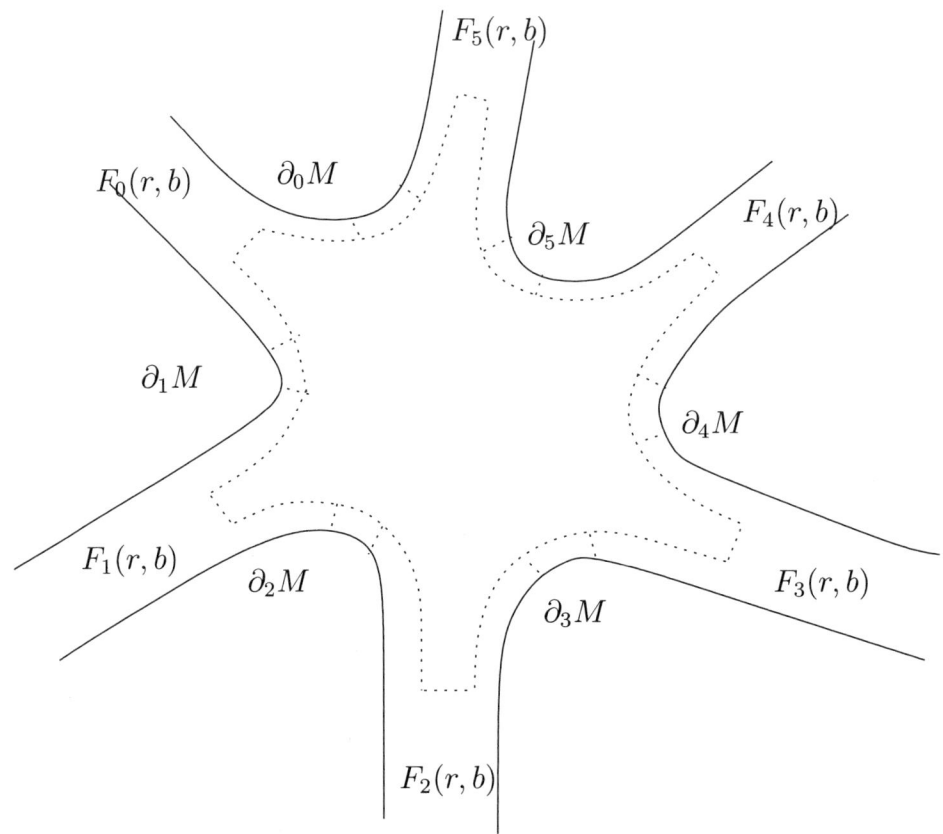

FIGURE 1. The manifold M

For $r \geq 0$ and $b \leq \frac{1}{4}$ we define an open covering $\mathcal{U}(r,b) = \{\mathcal{U}_k\}_{k \in J}$ of M as follows. Let J be the union of $\mathbb{Z}/6$ with a one-element set $\{\clubsuit\}$. For $k \in \mathbb{Z}/6$ let $\mathcal{U}_k := F_k(r,b)$ and let $\mathcal{U}_\clubsuit := M \setminus \overline{F(r+1, b/2)}$.

For $r \geq 0$ let $M_r := M \setminus \overline{F(r,0)}$.

The connected components of ∂M are labelled $\partial_k M$, $k \in \mathbb{Z}/6$, in such a way that $\partial_k M \cap F_k(r,b) \subset \{x_2^k = 0\}$ and $\partial_{k+1} M \cap F_k(r,b) \subset \{x_2^k = 1\}$.

The manifold M can be embedded diffeomorphically into $\mathbb{R}^2$, even with a diffeomorphism that is an isometry on the complement of M_r for some $r > 0$. The image of the embedding is illustrated by figure 1.

Now we define the Dirac operator. Choose a spin structure on M and fix $d \in \mathbb{N}$. Let S be the spinor bundle endowed with the Levi-Cività connection and a parallel hermitian metric. Let E be the graded vector bundle $S \otimes ((\mathbb{C}^+)^d \oplus (\mathbb{C}^-)^d)$. The hermitian metric on S and the standard hermitian product on $\mathbb{C}^d$ induce a hermitian metric $\langle \cdot, \cdot \rangle$ on E.

Furthermore there are connections on E and its dual E^* induced by the Clifford connection on S and the de Rham differential. The Dirac operator associated to the Dirac bundle E is denoted by $\partial\!\!\!/_E$.

The oriented orthonormal frames $((-1)^k dx_1^k, (-1)^k dx_2^k)$ of $TM|_{F_k(0,\frac{1}{4})}$ patch together to an oriented orthonormal frame (e_1, e_2) of $TM|_F$. The Clifford multiplication $c: TM \to \operatorname{End} E$ induces an even parallel endomorphism

$$I = c(e_2)c(e_1) : E|_F \to E|_F$$

defining a skew-hermitian form

$$E|_F \times E|_F \to \mathbb{C}, \ (x,y) \mapsto \langle x, Iy \rangle .$$

Since the holonomy of $TM|_F$ is $4\pi\mathbb{Z}$ (measured with respect to any trivialization of TM on the whole of M), there are nonvanishing parallel sections of $S^+|_F$ and $S^-|_F$. We choose a parallel unit section s of $S^+|_F$ and fix once and for all the trivialization of $E|_F$ defined for $x \in F$ by

$$E_x^+ = S_x^+ \otimes (\mathbb{C}^+)^d \oplus S_x^- \otimes (\mathbb{C}^-)^d \to (\mathbb{C}^+)^d \oplus (\mathbb{C}^+)^d ,$$

$$(s(x) \otimes v) \oplus (ic(e_1)s(x) \otimes w) \mapsto (v, w) ,$$

and

$$E_x^- = S_x^- \otimes (\mathbb{C}^+)^d \oplus S_x^+ \otimes (\mathbb{C}^-)^d \to (\mathbb{C}^-)^d \oplus (\mathbb{C}^-)^d ,$$

$$(ic(e_1)s(x) \otimes v) \oplus (s(x) \otimes w) \mapsto (v, w) .$$

With respect to this trivialization the endomorphism $I|_{E^+}$ corresponds to

$$I_0 := \begin{pmatrix} i & 0 \\ 0 & -i \end{pmatrix} \in M_{2d}(\mathbb{C})$$

and $I|_{E^-}$ to $-I_0$.

1.2. The family index theorem

In order to give a motivation for the definitions in the subsequent sections we sketch the corresponding family index theorem. The presentation follows [**BK**].

Let $\mathbb{C}^{2d}$ be endowed with the standard hermitian product $\langle \, , \, \rangle$ and the skew-hermitian form $(x,y) \mapsto \langle x, I_0 y \rangle$.

Let B be a compact space and let (L_0, L_1, L_2) be a triple of pairwise transverse Lagrangian subbundles of the trivial bundle $B \times \mathbb{C}^{2d}$. For any $b \in B$ the Lagrangian subspaces $L_i(b) \subset \mathbb{C}^{2d}$ define parallel Lagrangian subbundles of $E^+|_F$ via the trivialization fixed in the previous section.

Let $D^+(b)$ be the Dirac operator associated to E with

$$\operatorname{dom} D^+(b) := \{s \in C_c^\infty(M, E^+) \mid s(x) \in L_i(b) \text{ for } x \in \partial_i M \cup \partial_{i+3} M, \ i = 0, 1, 2\} .$$

It turns out that for any $b \in B$ the kernel and cokernel of the closure of $D^+(b)$ are finite dimensional and that the family $\{D^+(b)\}_{b \in B}$ has a well-defined index in $K^0(B)$, which equals the generalized Maslov index of (L_0, L_1, L_2) defined as follows.

The triple (L_0, L_1, L_2) induces a nondegenerate hermitian form h on L_0: Let $v = v_1 + v_2$, $w = w_1 + w_2 \in L_0(b)$ with $v_1, w_1 \in L_1(b)$; $v_2, w_2 \in L_2(b)$, then

$$h_b(v, w) := \langle v_2, I_0 w_1 \rangle .$$

The generalized Maslov index is the element $[L_0^+] - [L_0^-] \in K^0(B)$ where L_0^+ and L_0^- are subbundles of L_0 with $L_0^+ \oplus L_0^- = L_0$ and such that h is positive on L_0^+ and negative on L_0^-.

If B is a manifold and the bundles are smooth, then by a generalization of the Atiyah-Patodi-Singer index theorem the Chern character of the index bundle can be expressed in terms of η-forms; the local term vanishes since $\text{ch}(E/S) = 0$.

We outline the definition of the η-forms.

Let ∂ be the differentiation operator on $C^\infty([0,1], \mathbb{C}^{2d})$.

For $i \neq j$ and any $b \in B$ the operator $D_I(b) := I_0 \partial$ with domain
$$\text{dom}\, D_I(b) := \{s \in C^\infty([0,1], \mathbb{C}^{2d}) \mid s(0) \in L_i(b),\ s(1) \in L_j(b)\}$$
is essentially selfadjoint and its closure has a bounded inverse on $L^2([0,1], \mathbb{C}^{2d})$.

There is a family of rescaled superconnections A_t^I associated to the family of operators σD_I, where σ is a formal parameter of degree 1 with $\sigma^2 = 1$.

The η-form
$$\eta(L_i, L_j) := \frac{1}{\sqrt{\pi}} \int_0^\infty t^{-\frac{1}{2}} \text{Tr}_\sigma D_I e^{-(A_t^I)^2} dt \in \Omega^*(B) ,$$
with $\text{Tr}_\sigma(a + \sigma b) := \text{Tr}(a)$, is well-defined.

The statement of the index theorem is:
$$\text{ch}(\text{ind}\, D^+) = [\eta(L_0, L_1) + \eta(L_1, L_2) + \eta(L_2, L_0)] \in H^*_{dR}(B) .$$

1.3. The algebra of differential forms

In this section we study the noncommutative analogues of the algebra $\Omega^*(B)$, the Chern character and the de Rham cohomology $H^*_{dR}(B)$.

1.3.1. The universal graded differential algebra.
Let $\mathcal{B}$ be an involutive locally m-convex Fréchet algebra with unit. In particular, the multiplication $\mathcal{B} \times \mathcal{B} \to \mathcal{B}$, $(a,b) \mapsto ab$ is continuous, the group of invertible elements $\text{Gl}(\mathcal{B})$ in $\mathcal{B}$ is open and the map $\text{Gl}(\mathcal{B}) \to \text{Gl}(\mathcal{B})$, $a \mapsto a^{-1}$ is continuous [**Ma**]. In this section we recall the definition of the topological universal graded differential algebra $\hat{\Omega}_*\mathcal{B}$ and collect its main properties [**Ka**] [**CQ**].

We write $\otimes_\pi$ for the completed projective tensor product.

Let
$$\hat{\Omega}_k\mathcal{B} := \mathcal{B} \otimes_\pi (\otimes_\pi^k(\mathcal{B}/\mathbb{C}))$$
and
$$\hat{\Omega}_*\mathcal{B} := \prod_{k=0}^\infty \hat{\Omega}_k\mathcal{B} .$$

With the following structures $\hat{\Omega}_*\mathcal{B}$ is an involutive Fréchet locally m-convex graded differential algebra:

Product: There is a graded continuous product on $\hat{\Omega}_*\mathcal{B}$ defined for elementary tensors by
$$(b_0 \otimes b_1 \otimes \ldots \otimes b_k)(b_{k+1} \otimes b_{k+2} \otimes \ldots b_n) := \sum_{j=0}^k (-1)^{k-j} (b_0 \otimes b_1 \otimes \ldots \otimes b_j b_{j+1} \otimes b_{j+2} \otimes \ldots b_n) .$$

1.3. THE ALGEBRA OF DIFFERENTIAL FORMS

Differential: There is a continuous differential d of degree one on the graded algebra $\hat{\Omega}_*\mathcal{B}$ defined on elementary tensors by

$$\mathrm{d}(b_0 \otimes b_1 \otimes \ldots \otimes b_k) := 1 \otimes b_0 \otimes b_1 \otimes \ldots \otimes b_k .$$

It satisfies the graded Leibniz rule: For $\alpha \in \hat{\Omega}_k\mathcal{B}$ and $\beta \in \hat{\Omega}_*\mathcal{B}$

$$\mathrm{d}(\alpha\beta) = (\mathrm{d}\,\alpha)\beta + (-1)^k \alpha(\mathrm{d}\,\beta) .$$

Furthermore in $\hat{\Omega}_k\mathcal{B}$

$$b_0 \otimes b_1 \otimes \ldots \otimes b_k = b_0 \,\mathrm{d}\, b_1 \,\mathrm{d}\, b_2 \ldots \mathrm{d}\, b_k .$$

If $\mathcal{B}$ is a Banach algebra, then d is a map of norm one.

Involution: We extend the $*$-operation on $\mathcal{B}$ to a continuous involution on $\hat{\Omega}_*\mathcal{B}$ by setting

$$(b_0 \otimes b_1 \otimes \ldots \otimes b_k)^* := (1 \otimes b_k^* \otimes b_{k-1}^* \otimes \ldots \otimes b_1^*)b_0^*$$

or equivalently

$$(b_0 \,\mathrm{d}\, b_1 \,\mathrm{d}\, b_2 \ldots \mathrm{d}\, b_k)^* = (\mathrm{d}\, b_k^* \,\mathrm{d}\, b_{k-1}^* \ldots \mathrm{d}\, b_1^*)b_0^* .$$

For $\omega_1, \omega_2 \in \hat{\Omega}_*\mathcal{B}$

$$(\omega_1 \omega_2)^* = \omega_2^* \omega_1^* ,$$

and for $\omega \in \hat{\Omega}_k\mathcal{B}$

$$(\mathrm{d}\,\omega)^* = (-1)^k \,\mathrm{d}(\omega^*) .$$

Let

$$\hat{\Omega}_{\leq m}\mathcal{B} := \hat{\Omega}_*\mathcal{B} / \prod_{k=m+1}^{\infty} \hat{\Omega}_k\mathcal{B} .$$

The above structures are well-defined on $\hat{\Omega}_{\leq m}\mathcal{B}$ as well.

We identify $\hat{\Omega}_{\leq m}\mathcal{B}$ as a graded vector space with the subspace

$$\{\omega \in \hat{\Omega}_*\mathcal{B} \mid \omega^n = 0 \text{ for } n > m\} \subset \hat{\Omega}_*\mathcal{B} ,$$

where ω^n denotes the homogeneous part of degree n of $\omega \in \hat{\Omega}_*\mathcal{B}$.

The Fréchet space $\overline{[\hat{\Omega}_*\mathcal{B}, \hat{\Omega}_*\mathcal{B}]_s}$, generated by the supercommutators in $\hat{\Omega}_*\mathcal{B}$, is preserved by d by Leibniz rule. It follows that $(\hat{\Omega}_*\mathcal{B}/\overline{[\hat{\Omega}_*\mathcal{B}, \hat{\Omega}_*\mathcal{B}]_s}, \mathrm{d})$ is a complex.

DEFINITION 1.3.1. *The* DE RHAM HOMOLOGY *of $\mathcal{B}$ is*

$$H_*^{dR}(\mathcal{B}) := H_*(\hat{\Omega}_*\mathcal{B}/\overline{[\hat{\Omega}_*\mathcal{B}, \hat{\Omega}_*\mathcal{B}]_s}, \mathrm{d}) .$$

On the right hand side we take the topological homology, i.e. we quotient out the closure of the range of d *in order to obtain a Hausdorff space.*

The map d induces maps $\mathrm{d} : (\hat{\Omega}_*\mathcal{B})^n \to (\hat{\Omega}_*\mathcal{B})^n$ and $\mathrm{d} : M_n(\hat{\Omega}_*\mathcal{B}) \to M_n(\hat{\Omega}_*\mathcal{B})$, $A \mapsto \mathrm{d}(A)$ by applying d componentwise. Note the difference between $\mathrm{d}\, A = \mathrm{d} \circ A$ and $\mathrm{d}(A)$. Sometimes we write $(\mathrm{d}\, A)$ for $\mathrm{d}(A)$.

For $A \in M_n(\hat{\Omega}_k\mathcal{B})$

$$\mathrm{d}\, A = (\mathrm{d}\, A) + (-1)^k A\,\mathrm{d} .$$

The trace

$$\mathrm{tr} : M_n(\hat{\Omega}_*\mathcal{B}) \to \hat{\Omega}_*\mathcal{B}/\overline{[\hat{\Omega}_*\mathcal{B}, \hat{\Omega}_*\mathcal{B}]_s}$$

is defined by adding up the diagonal elements. It vanishes on supercommutators.

In the following definition and proposition a projection is not assumed to be selfadjoint.

DEFINITION 1.3.2. Let $P \in M_n(\mathcal{B})$ be a projection. Then
$$\text{ch}(P) := \sum_{k=0}^{\infty} (-1)^k \frac{1}{k!} \text{tr } P(\text{d}(P))^{2k} \in \hat{\Omega}_*\mathcal{B}/\overline{[\hat{\Omega}_*\mathcal{B}, \hat{\Omega}_*\mathcal{B}]_s}$$

is the CHERN CHARACTER FORM of P.

PROPOSITION 1.3.3. (1) *The Chern character form is closed.*
(2) *If $P : [0,1] \to M_n(\mathcal{B})$ is a differentiable path of projections, then $\text{ch}(P_1) - \text{ch}(P_0)$ is exact.*
(3) *If $\mathcal{B}$ is a local Banach algebra* [**Bl**]*, then the Chern character form induces a homomorphism*
$$\text{ch} : K_0(\mathcal{B}) \to H_*^{dR}(\mathcal{B}), \; \text{ch}([P] - [Q]) := \text{ch}(P) - \text{ch}(Q) \; ,$$

called the CHERN CHARACTER.

PROOF. First note that for a projection $P \in M_n(\mathcal{B})$
$$0 = \text{d}((1-P)P) = (1-P)(\text{d}\,P) - (\text{d}\,P)P \; ,$$

hence $(1-P)(\text{d}\,P) = (\text{d}\,P)P$ and $P(\text{d}\,P)P = 0$. Therefore
$$P(\text{d}\,P)^2 = (\text{d}\,P)^2 P \; .$$

Analogous formulas hold for the derivative of a path of projections.
(1) follows from
$$\begin{aligned}\text{d tr } P(\text{d}\,P)^{2k} &= \text{tr}(\text{d}\,P)^{2k+1} \\ &= \text{tr}(1-P)(\text{d}\,P)^{2k+1}(1-P) + \text{tr } P(\text{d}\,P)^{2k+1}P \\ &= 0 \; .\end{aligned}$$

(2) We have that
$$\begin{aligned}(\text{tr } P(\text{d}\,P)^{2k})' &= \text{tr } P'(\text{d}\,P)^{2k} + \text{tr } P((\text{d}\,P)^{2k})' \\ &= \sum_{i=0}^{2k-1} \text{tr } P(\text{d}\,P)^i(\text{d}\,P')(\text{d}\,P)^{2k-i-1} \; .\end{aligned}$$

For i even
$$\begin{aligned}\text{tr } &P(\text{d}\,P)^i(\text{d}\,P')(\text{d}\,P)^{2k-i-1} \\ &= \text{tr}(\text{d}\,P)^i(\text{d}(PP'))(\text{d}\,P)^{2k-i-1} - \text{tr}(\text{d}\,P)^i(\text{d}\,P)P'(\text{d}\,P)^{2k-i-1} \\ &= \text{tr}(\text{d}\,P)^i(\text{d}(PP'))(\text{d}\,P)^{2k-i-1} \\ &= \text{d tr } P(\text{d}\,P)^{i-1}(\text{d}(PP'))(\text{d}\,P)^{2k-i-1} \; .\end{aligned}$$

For i odd the argument is similar.
(3) follows from (1) and (2). □

1.3.2. Supercalculus. Let $\mathcal{B}$ be as before.

In this section all spaces and tensor products are $\mathbb{Z}/2$-graded. If no grading is specified we assume the grading to be trivial.

Let $V = V^+ \oplus V^-$ be a $\mathbb{Z}/2$-graded complex vector space with $\dim V^+ = m$, $\dim V^- = n$ and consider $\hat{\Omega}_*\mathcal{B}$ as a $\mathbb{Z}/2$-graded space with the grading induced by the form degree.

The space $V \otimes \hat{\Omega}_*\mathcal{B}$ is a free $\mathbb{Z}/2$-graded right $\hat{\Omega}_*\mathcal{B}$-module. It is furthermore a left supermodule of the superalgebra $\mathrm{End}(V) \otimes \hat{\Omega}_*\mathcal{B}$. (Note that our setting differs from the corresponding one in [**BGV**] since we deal with right $\hat{\Omega}_*\mathcal{B}$-modules. This leads to different signs.)

The supertrace $\mathrm{tr}_s : \mathrm{End}(V) \to \mathbb{C}$ extends to a supertrace

$$\mathrm{tr}_s : \mathrm{End}(V) \otimes \hat{\Omega}_*\mathcal{B} \to \hat{\Omega}_*\mathcal{B}/\overline{[\hat{\Omega}_*\mathcal{B}, \hat{\Omega}_*\mathcal{B}]_s}$$

$$\mathrm{tr}_s(T \otimes \omega) := \mathrm{tr}_s(T)\omega \ .$$

By quotienting out the supercommutator we ensure that $\mathrm{tr}_s([T_1, T_2]_s) = 0$ for $T_1, T_2 \in \mathrm{End}(V) \otimes \hat{\Omega}_*\mathcal{B}$.

The differential d acts on elements of $V \otimes \hat{\Omega}_*\mathcal{B}$ resp. $\mathrm{End}(V) \otimes \hat{\Omega}_*\mathcal{B}$ by

$$\mathrm{d}(A \otimes \omega) = (-1)^{\deg A} A \otimes \mathrm{d}\omega$$

for $A \in V^\pm$ resp. $A \in \mathrm{End}^\pm(V)$ and $\omega \in \hat{\Omega}_*\mathcal{B}$.

Though the differential d is not a right $\hat{\Omega}_*\mathcal{B}$-module map, the supercommutator $[\mathrm{d}, T]_s$ with $T \in \mathrm{End}(V) \otimes \hat{\Omega}_*\mathcal{B}$ is in $\mathrm{End}(V) \otimes \hat{\Omega}_*\mathcal{B}$; namely

$$[\mathrm{d}, T]_s = \mathrm{d}(T) \ .$$

Since $\mathrm{tr}_s(A \otimes \omega) = 0$ for $A \in \mathrm{End}^-(V)$, we have that

$$\mathrm{tr}_s[\mathrm{d}, T]_s = \mathrm{tr}_s \mathrm{d}(T) = \mathrm{d}\, \mathrm{tr}_s T$$

for $T \in \mathrm{End}(V) \otimes \hat{\Omega}_*\mathcal{B}$.

See §5.2.3, §5.2.4, §5.2.5 for notions of Hilbert-Schmidt operators, trace class operators and adjointable operators used in the following.

If M is a complete Riemannian manifold and T is a trace class operator on $L^2(M, V \otimes \hat{\Omega}_{\leq\mu}\mathcal{B})$, then

$$\mathrm{Tr}_s T := \int_M \mathrm{tr}_s \overline{R}(T)(x)\, dx \in \hat{\Omega}_{\leq\mu}\mathcal{B}/\overline{[\hat{\Omega}_{\leq\mu}\mathcal{B}, \hat{\Omega}_{\leq\mu}\mathcal{B}]_s} \ .$$

Note that

$$\mathrm{Tr}_s[A, B]_s = 0$$

if A is an adjointable operator and B a trace class operator or if A, B are Hilbert-Schmidt operators.

1.3.3. The algebras $\mathcal{A}_\infty$ and $\hat{\Omega}_*\mathcal{A}_\infty$.
Let $(\mathcal{A}_j, \iota_{j+1,j} : \mathcal{A}_{j+1} \to \mathcal{A}_j)_{j \in \mathbb{N}_0}$ be a projective system of involutive Banach algebras with unit satisfying the following conditions (see [**Lo3**], §2.1):

- The algebra $\mathcal{A} := \mathcal{A}_0$ is a C^*-algebra.
- For any $j \in \mathbb{N}_0$ the map $\iota_{j+1,j} : \mathcal{A}_{j+1} \to \mathcal{A}_j$ is injective.
- For any $j \in \mathbb{N}_0$ the map $\iota_j : \mathcal{A}_\infty := \varprojlim_i \mathcal{A}_i \to \mathcal{A}_j$ has dense range.

- For any $j \in \mathbb{N}_0$ the algebra $\mathcal{A}_j$ is stable with respect to the holomorphic functional calculus in $\mathcal{A}$.

The projective limit $\mathcal{A}_\infty$ is an involutive locally m-convex Fréchet algebra with unit.

The motivating example is $\mathcal{A}_j = C^j(B)$ for a closed smooth manifold B.

PROPOSITION 1.3.4. *For $j \in \mathbb{N}_0$ and $n \in \mathbb{N}$ we have:*
(1) *The map $\iota_j : \mathcal{A}_\infty \to \mathcal{A}_j$ is injective.*
(2) *The algebras $M_n(\mathcal{A}_\infty)$ and $M_n(\mathcal{A}_j)$ are stable with respect to the holomorphic functional calculus in $M_n(\mathcal{A})$.*
(3) *The map $\iota_{0*} : K_0(\mathcal{A}_\infty) \to K_0(\mathcal{A})$ is an isomorphism.*

PROOF. (1) follows immediately.
(2) follows from [**Bo**], Prop. A.2.2.
(3) follows from [**Bo**], Th. A.2.1. □

The projective system $(\mathcal{A}_j, \iota_{j+1,j})_{j \in \mathbb{N}_0}$ induces two projective systems of involutive graded differential Fréchet algebras.

One of them is given by the maps
$$\iota_{j+1,j*} : \hat{\Omega}_* \mathcal{A}_{j+1} \to \hat{\Omega}_* \mathcal{A}_j \ .$$

Furthermore $(\hat{\Omega}_{\leq m} \mathcal{A}_j)_{m,j \in \mathbb{N}}$ is a projective system of involutive Banach graded differential algebras.

Their limits coincide:

The inclusion $\hat{\Omega}_{\leq m} \mathcal{A}_j \to \hat{\Omega}_* \mathcal{A}_j$ is left inverse to the projection $\hat{\Omega}_* \mathcal{A}_j \to \hat{\Omega}_{\leq m} \mathcal{A}_j$. The induced maps between the projective limits are inverse to each other. It follows that
$$\varprojlim_{j} \hat{\Omega}_* \mathcal{A}_j \cong \varprojlim_{j,m} \hat{\Omega}_{\leq m} \mathcal{A}_j \ .$$

Furthermore the inclusions $\iota_{j*} : \hat{\Omega}_* \mathcal{A}_\infty \to \hat{\Omega}_* \mathcal{A}_j$ induce a map
$$\iota_* : \hat{\Omega}_* \mathcal{A}_\infty \to \varprojlim_{j} \hat{\Omega}_* \mathcal{A}_j \ .$$

PROPOSITION 1.3.5. *There are the following canonical isomorphisms of involutive Fréchet locally m-convex graded differential algebras:*
(1) $\hat{\Omega}_* \mathcal{A}_\infty \cong \varprojlim_{j} \hat{\Omega}_* \mathcal{A}_j \cong \varprojlim_{j,m} \hat{\Omega}_{\leq m} \mathcal{A}_j \ .$

There are the following canonical isomorphisms of graded Fréchet spaces:
(2) $\hat{\Omega}_* \mathcal{A}_\infty / \overline{[\hat{\Omega}_* \mathcal{A}_\infty, \hat{\Omega}_* \mathcal{A}_\infty]_s} \cong \varprojlim_{j,m} \hat{\Omega}_{\leq m} \mathcal{A}_j / \overline{[\hat{\Omega}_{\leq m} \mathcal{A}_j, \hat{\Omega}_{\leq m} \mathcal{A}_j]_s} \ ,$
(3) $H_*^{dR}(\mathcal{A}_\infty) \cong \varprojlim_{j} H_*^{dR}(\mathcal{A}_j) \cong \varprojlim_{j,m} H_*(\hat{\Omega}_{\leq m} \mathcal{A}_j / \overline{[\hat{\Omega}_{\leq m} \mathcal{A}_j, \hat{\Omega}_{\leq m} \mathcal{A}_j]_s}, d) \ .$

PROOF. (1) It is enough to prove that the right hand side and the left hand side are isomorphic as topological vector spaces. This follows from the fact that projective limits and projective tensor products commute ([**Kö**], 41.6).

(2) and (3) follow from the three technical lemmas below. □

The importance of this proposition for our purposes is the following: Since the analysis is easier on Banach spaces than on Fréchet spaces we will prove the index theorem in $H_*(\hat{\Omega}_{\leq m}\mathcal{A}_j/\overline{[\hat{\Omega}_{\leq m}\mathcal{A}_j,\hat{\Omega}_{\leq m}\mathcal{A}_j]_s}, d)$, making sure that the expressions in the index theorem behave well under the projective limit. By the proposition this will prove the index theorem in $H_*^{dR}(\mathcal{A}_\infty)$.

LEMMA 1.3.6. *Let $(V_n, f_{n+1,n} : V_{n+1} \to V_n)_{n \in \mathbb{N}}$ be a projective system of Banach spaces with projective limit V_∞. Let $(A_n, f_{n+1,n}|_{A_{n+1}})_{n \in \mathbb{N}}$ be a projective subsystem of sets such that for any $n \in \mathbb{N}$ the range of the induced map $f_n : A_\infty := \varprojlim_n A_n \to A_n$ is dense in A_n. Then*

$$\varprojlim_n \overline{A_n} = \overline{\varprojlim_n A_n} \ .$$

Hence a subset S_1 of V_∞ is dense in $S_2 \subset V_\infty$ if and only if $f_n S_1 \subset V_n$ is dense in $f_n S_2 \subset V_n$ for any $n \in \mathbb{N}$.

PROOF. Without loss of generality we may assume that the maps $f_{n+1,n}$ have norm less than or equal to one. Let $a \in \varprojlim_n \overline{A_n}$. Then for any $n \in \mathbb{N}$ there is $b_n \in A_\infty$ with $|f_n(b_n) - f_n(a)| \leq \frac{1}{n}$. The sequence $(b_n)_{n \in \mathbb{N}}$ converges in V_∞ to a. Hence $\overline{A_\infty} = \varprojlim_n \overline{A_n}$. □

LEMMA 1.3.7. (1) *Let $(V_n, f_{n+1,n})_{n \in \mathbb{N}}$, $(W_n, g_{n+1,n})_{n \in \mathbb{N}}$ be projective systems of Banach spaces. Let $(d_n : V_n \to W_n)_{n \in \mathbb{N}}$ be a morphism of projective systems with induced map $d_\infty : V_\infty \to W_\infty$. Then*

$$\operatorname{Ker} d_\infty = \varprojlim_n \operatorname{Ker} d_n \ .$$

(2) *In the situation of (1) assume furthermore that the induced maps $f_n : V_\infty \to V_n$ and $g_n : W_\infty \to W_n$ have dense range for all $n \in \mathbb{N}$. Then*

$$\overline{\operatorname{Ran} d_\infty} = \varprojlim_n \overline{\operatorname{Ran} d_n} \ .$$

(3) *Let $(A_n, f_{n+1,n})_{n \in \mathbb{N}}$ be a projective system of Banach algebras such that for any $n \in \mathbb{N}$ the range of $f_n : A_\infty := \varprojlim_n A_n \to A_n$ is dense in A_n. Then*

$$\overline{[A_\infty, A_\infty]} = \varprojlim_n \overline{[A_n, A_n]} \ .$$

PROOF. (1) Let $x \in \operatorname{Ker} d_\infty$. Then $d_n f_n(x) = g_n d_\infty(x) = 0$.
Conversely if $d_n f_n x = 0$ for all $n \in \mathbb{N}$, then by definition $d_\infty x = 0$.
(2) follows from the previous lemma since $g_n \operatorname{Ran} d_\infty$ is dense in $\overline{\operatorname{Ran} d_n}$.
(3) follows from the previous lemma since $f_n[A_\infty, A_\infty]$ is dense in $\overline{[A_n, A_n]}$ for all $n \in \mathbb{N}$. □

LEMMA 1.3.8. *Let $(V_n, f_{n+1,n})_{n \in \mathbb{N}}$ be a projective system of Banach spaces and let $(A_n, f_{n+1,n}|_{A_{n+1}})_{n \in \mathbb{N}}$ be a projective subsystem such that A_n is a closed subspace of V_n for all $n \in \mathbb{N}$.*

Let $V_\infty := \varprojlim_n V_n$ and $A_\infty := \varprojlim_n A_n$ and assume furthermore that the image of A_{n+1} is dense in A_n for any $n \in \mathbb{N}$.

Then, canonically,

$$V_\infty / A_\infty \cong \varprojlim_n V_n / A_n \ .$$

PROOF. For $n \in \mathbb{N}$ let $f_n : V_\infty \to V_n$ be the induced map. We prove that the map
$$f_\infty : V_\infty/A_\infty \to \varprojlim_n V_n/A_n$$
is an isomorphism:

For injectivity let $[v] \in V_\infty/A_\infty$ with $f_\infty[v] = 0$. Then $f_n v \in A_n$ for all $n \in \mathbb{N}$, so $v \in A_\infty$, thus $[v] = 0$.

In order to prove surjectivity we assume that the norms of the maps $f_{n+1,n}$ are less than or equal to one, which can be obtained by rescaling the norms inductively.

Let $v \in \varprojlim_n V_n/A_n$. Choose inductively $v_n \in V_n$ such that $[v_n] \in V_n/A_n$ is the image of v with respect to the map $\varprojlim_n V_n/A_n \to V_n/A_n$ and such that
$$|f_{n+1,n} v_{n+1} - v_n| \leq n^{-1} .$$
(Here we use that $f_{n+1,n} A_{n+1}$ is dense in A_n.)

For $n \geq j$ let $f_{n,j} : V_n \to V_j$ be the induced map. For any $j \in \mathbb{N}$ the sequence $(f_{n,j}(v_n))_{n \geq j}$ converges to an element $\tilde{v}_j \in V_j$. We have that $f_{j+1,j} \tilde{v}_{j+1} = \tilde{v}_j$, hence there is an element $\tilde{v} \in V_\infty$ with $f_j \tilde{v} = \tilde{v}_j$. Since $[\tilde{v}_j] = [v_j] \in V_j/A_j$, we have that $f_\infty[\tilde{v}] = v$.

□

1.4. Lagrangian projections

Let $\mathcal{A}$ be a unital C^*-algebra.

In this section we define and study the analogues of Lagrangian subbundles and of the Maslov index bundle.

Let $n \in \mathbb{N}$.

DEFINITION 1.4.1. *Two selfadjoint projections $P_1, P_2 \in M_n(\mathcal{A})$ are called* TRANSVERSE *if*
$$\operatorname{Ran} P_1 \oplus \operatorname{Ran} P_2 = \mathcal{A}^n .$$

We will often use the following transversality criterion: Two selfadjoint projections P_1, P_2 are transverse if and only there exist $a, b \in \operatorname{Gl}(\mathcal{A})$ such that $aP_1 + bP_2 \in M_n(\mathcal{A})$ is invertible. This is equivalent to the invertibility of $aP_1 + bP_2$ for any $a, b \in \operatorname{Gl}(\mathcal{A})$.

If $\mathcal{A} = C(B)$ for a compact space B, then the transversality of two projections is equivalent to the transversality of the corresponding subbundles.

1.4.1. Definition and properties. Let $\mathcal{A}^{2n}$ be endowed with the standard $\mathcal{A}$-valued scalar product and let
$$I_0 = \begin{pmatrix} i & 0 \\ 0 & -i \end{pmatrix} : \mathcal{A}^n \oplus \mathcal{A}^n \to \mathcal{A}^n \oplus \mathcal{A}^n .$$

DEFINITION 1.4.2. *A* LAGRANGIAN PROJECTION ON $\mathcal{A}^{2n}$ *is a selfadjoint projection $P \in M_{2n}(\mathcal{A})$ with*
$$PI_0 = I_0(1-P) .$$

If $\mathcal{A} = C(B)$ for some compact space B, then the Lagrangian projections are in one-to-one correspondence with the subbundles of $B \times \mathbb{C}^{2n}$ that are Lagrangian with respect to the skew-hermitian form induced by I_0.

Note furthermore that any Lagrangian projection P on $\mathcal{A}^{2n}$ is also a Lagrangian projection on the Hilbert $M_n(\mathcal{A})$-module $M_n(\mathcal{A})^2$.

Denote
$$P_s := \tfrac{1}{2}\begin{pmatrix} 1 & 1 \\ 1 & 1 \end{pmatrix}.$$

LEMMA 1.4.3. (1) *For every Lagrangian projection P on $\mathcal{A}^{2n}$ there is a unitary $p \in M_n(\mathcal{A})$ such that*
$$P = \tfrac{1}{2}\begin{pmatrix} 1 & p^* \\ p & 1 \end{pmatrix}.$$

(2) *For every Lagrangian projection P on $\mathcal{A}^{2n}$ the unitary*
$$U = \begin{pmatrix} 1 & 0 \\ 0 & p^* \end{pmatrix} \in M_{2n}(\mathcal{A})$$
with p as in (1) fulfills $UI_0 = I_0 U$ and $UPU^* = P_s$.

PROOF. (1) Since P is selfadjoint, there are $a, b, c \in M_n(\mathcal{A})$ with $a = a^*, c = c^*$ such that $P = \begin{pmatrix} a & b \\ b^* & c \end{pmatrix}$. From $PI_0 = I_0(1-P)$ it follows that
$$\begin{pmatrix} ia & -ib \\ ib^* & -ic \end{pmatrix} = \begin{pmatrix} i(1-a) & -ib \\ ib^* & -i(1-c) \end{pmatrix},$$
thus $a = c = \tfrac{1}{2}$. Furthermore from $P^2 = P$ it follows that $2b$ is unitary.

(2) is clear. □

Let $\mathcal{A}_\infty$ be as in the previous section.

LEMMA 1.4.4. *Let $P \in M_{2n}(\mathcal{A}_\infty)$ be a Lagrangian projection of $\mathcal{A}^{2n}$ transverse to P_s. Let $\tilde P \in M_{2n}(\mathbb{C})$ be a complex Lagrangian projection. Then for every $0 < \varepsilon_1 < \varepsilon_2$ there is a smooth path of unitaries $U : [0, \varepsilon_2] \to M_{2n}(\mathcal{A}_\infty)$ such that*

(1) $U(0)PU(0)^* = \tilde P$,
(2) U *equals 1 on a neighborhood of ε_2,*
(3) U *is constant on $[0, \varepsilon_1]$,*
(4) U *is diagonal with respect to the decomposition $\mathcal{A}^{2n} = \mathcal{A}^n \oplus \mathcal{A}^n$.*

Note that (4) implies that $UI_0 = I_0 U$.

PROOF. It is enough to prove the assertion for $\tilde P = P_s$.

For P let p be as in the previous lemma.

Since P and P_s are transverse, $P - P_s$ is invertible. It follows that $p - 1$ and $p^* - 1$ are invertible, so $\log p$ and $\log p^*$ are well-defined if we choose the complement of $[0, \infty)$ in $\mathbb{C}$ as a domain for the logarithm.

Let $\chi : [0, \varepsilon_2] \to [0, 1]$ be a smooth function with $\chi|_{[0,\varepsilon_1]} = 0$ and $\chi(t) = 1$ for $t \in [\tfrac{\varepsilon_2 - \varepsilon_1}{2}, \varepsilon_2]$. The smooth path of unitaries
$$\gamma : [0, \varepsilon_2] \to M_n(\mathcal{A}), \ \gamma(t) := \exp(2\pi i \chi(t) + (1 - \chi(t))\log p^*)$$
connects p^* with 1. Moreover $\gamma(t) \in M_n(\mathcal{A}_\infty)$ for all $t \in [0, \varepsilon_2]$ since $M_n(\mathcal{A}_\infty)$ is stable under holomorphic function calculus in $M_n(\mathcal{A})$ by Prop. 1.3.4.

Then
$$U := \begin{pmatrix} 1 & 0 \\ 0 & \gamma \end{pmatrix}$$
satisfies the conditions. □

1.4.2. The Maslov index. Let (P_0, P_1, P_2) be a triple of pairwise transverse Lagrangian projections on $\mathcal{A}^{2n}$. For $x \in \mathcal{A}^{2n}$ write $x = x_1 + x_2$ with $x_i \in \operatorname{Ran} P_i$, $i = 1, 2$.

The form
$$h : \operatorname{Ran} P_0 \times \operatorname{Ran} P_0 \to \mathcal{A}, \quad (x, y) \mapsto \langle x_2, I_0 y_1 \rangle$$
is hermitian and its radical vanishes [**Wa**]. Since $x_1 = P_1(P_1 + P_2)^{-1} x$ and $x_2 = P_2(P_1 + P_2)^{-1} x$, the corresponding selfadjoint matrix is
$$A := P_0(P_1 + P_2)^{-1} P_2 I_0 P_1 (P_1 + P_2)^{-1} P_0 \in M_{2n}(\mathcal{A}) .$$

Since
$$\begin{aligned} A &= P_0(P_1 + P_2)^{-1} P_2 I_0 P_1 (P_1 + P_2)^{-1} P_0 \\ &= P_0(P_1 + P_2)^{-1}(P_1 + P_2) I P_1 (P_1 + P_2)^{-1}(P_0 + P_2) \\ &= (P_0 + P_1) I P_1 (P_1 + P_2)^{-1}(P_0 + P_1) , \end{aligned}$$
the range of A is closed. Hence the hermitian form h is non-singular and defines an element in $K_0(\mathcal{A})$ [**Ros2**].

DEFINITION 1.4.5. *The* MASLOV INDEX $\tau(P_0, P_1, P_2) \in K_0(\mathcal{A})$ *of a triple of pairwise transverse Lagrangian projections* (P_0, P_1, P_2) *is the class of the hermitian form* h *in* $K_0(\mathcal{A})$.

We can express the Maslov index in terms of A as follows:
$$\tau(P_0, P_1, P_2) = [1_{\{x > 0\}}(A)] - [1_{\{x < 0\}}(A)] \in K_0(\mathcal{A}) .$$

Note that the Maslov index is invariant under even permutations and changes sign under odd permutations.

PROPOSITION 1.4.6. *Let* $P_i : [0, 1] \to M_{2n}(\mathcal{A})$, $i = 0, 1, 2$, *be continuous paths of Lagrangian projections such that* $P_i(t) - P_j(t)$ *is invertible for* $i \neq j$ *and all* $t \in [0, 1]$.

Then the Maslov index $\tau(P_0(t), P_1(t), P_2(t))$ *does not depend on* t.

PROOF. The selfadjoint element $A(t) \in M_{2n}(\mathcal{A})$ defined by $(P_0(t), P_1(t), P_2(t))$ as above depends continuously on t for all $t \in [0, 1]$. It follows that the projections $1_{\{x > 0\}}(A(t))$ and $1_{\{x < 0\}}(A(t))$ also depend continuously on t, thus their K-theory classes are constant. □

Let B be a compact space and let (P_0, P_1, P_2) be a triple of pairwise transverse Lagrangian projections in $M_{2n}(C(B))$. Let (L_0, L_1, L_2) be the corresponding triple of Lagrangian subbundles of $B \times \mathbb{C}^{2n}$. Then the Maslov index bundle $[L_0^+] - [L_0^-]$ defined in §1.2 corresponds to $\tau(P_0, P_1, P_2)$ under the canonical isomorphism $K^0(B) \cong K_0(C(B))$.

Now we study in some detail the Maslov index of a triple (P_0, P_1, P_2) with $P_0 = P_s$. The general case can be reduced to this case by Lemma 1.4.3.

The Cayley transform $a \mapsto \frac{a-i}{a+i}$, defined for selfadjoint $a \in M_n(\mathcal{A})$, yields a bijective map
$$a \mapsto P(a) := \tfrac{1}{2}\begin{pmatrix} 1 & \frac{a+i}{a-i} \\ \frac{a-i}{a+i} & 1 \end{pmatrix} .$$
from the space of selfadjoint elements in $M_n(\mathcal{A})$ to space of projections in $M_{2n}(\mathcal{A})$ transverse to P_s.

1.4. LAGRANGIAN PROJECTIONS

LEMMA 1.4.7. *Let $a_1, a_2 \in M_n(\mathcal{A})$ be selfadjoint. Then $P(a_1)$ and $P(a_2)$ are transverse projections if and only if $a_1 - a_2$ is invertible.*

PROOF. Let $U := \frac{1}{\sqrt{2}} \begin{pmatrix} 1 & 1 \\ 1 & -1 \end{pmatrix}$. Then

$$UP(a_j)U^* = (a_j^2 + 1)^{-1} \begin{pmatrix} a_j^2 & -ia_j \\ ia_j & 1 \end{pmatrix}.$$

Now $P(a_1)$ and $P(a_2)$ are transverse if and only if

$$(a_1^2 + 1)UP(a_1)U^* - (a_2^2 + 1)UP(a_2)U^*$$

is invertible and this is the case if and only if $a_1 - a_2$ is invertible. □

LEMMA 1.4.8. *Let (P_s, P_1, P_2) be a triple of pairwise transverse Lagrangian projections and let $a_1, a_2 \in M_n(\mathcal{A})$ be such that $P_i = P(a_i)$, $i = 1, 2$. Let $p^+ := 1_{\{x>0\}}(a_1 - a_2)$.*

(1) There are continuous paths $P_1, P_2 : [0,2] \to M_{2n}(\mathcal{A})$ of Lagrangian projections such that $P_s, P_1(t), P_2(t)$ are pairwise transverse for all $t \in [0,2]$ and such that $P_1(2) = P(2p^+ - 1)$ and $P_2(2) = P(1 - 2p^+)$.

(2)
$$\tau(P_s, P_1, P_2) = [p^+] - [1 - p^+].$$

PROOF. (1) For $t \in [0,1]$ define $a_1(t) := t(a_1 - a_2) + (1-t)a_1$ and $a_2(t) := t(a_2 - a_1) + (1-t)a_2$. Then $a_1(t) - a_2(t) = (1+t)(a_1 - a_2)$ is invertible, thus the projections $P(a_1(t))$ and $P(a_2(t))$ are transverse. Furthermore $a_1(0) = a_1$ and $a_1(1) = a_1 - a_2$, whereas $a_2(0) = a_2$ and $a_2(1) = a_2 - a_1$.

For $t \in [1,2]$ let $a_1(t)$ be a path of invertible selfadjoint elements with $a_1(1) = a_1 - a_2$ and $a_1(2) = p^+ - (1 - p^+)$ and let $a_2(t) = -a_1(t)$.

Then the paths $P(a_1(t))$ and $P(a_2(t))$ satisfy the conditions.

(2) From (1) and Prop. 1.4.6 it follows that $\tau(P_s, P_1, P_2) = \tau(P_s, P(2p^+ - 1), P(1 - 2p^+))$.

The Cayley transform of $2p^+ - 1$ is $i(2p^+ - 1)$. By computing the matrix A we see that $1_{\{x>0\}}(A) = p^+$ and $1_{\{x<0\}}(A) = (1 - p^+)$. □

CHAPTER 2

The Fredholm Operator and Its Index

2.1. The operator D on M

2.1.1. Definition of D. Now we come back to the geometric situation described in §1.1.

By taking the tensor product of the bundle E with the C^*-algebra $\mathcal{A}$ we obtain an $\mathcal{A}$-vector bundle [**MF**]. Furthermore we consider the bundle $E \otimes \hat{\Omega}_{\leq \mu}\mathcal{A}_i$, $i, \mu \in \mathbb{N}_0$, of right $\hat{\Omega}_{\leq \mu}\mathcal{A}_i$-modules. Keep in mind that E can be trivialized on M via a global orthonormal frame. Thus no theory of Banach space bundles is needed in this context.

The hermitian metric on E extends to an $\mathcal{A}$-valued scalar product $\langle \cdot, \cdot \rangle$ on $E \otimes \mathcal{A}$ and to an $\hat{\Omega}_{\leq \mu}\mathcal{A}_i$-valued non-degenerated product on $E \otimes \hat{\Omega}_{\leq \mu}\mathcal{A}_i$ (see §5.2.3 for this notion). Furthermore parallel transport is defined on $E \otimes \mathcal{A}$ resp. $E \otimes \hat{\Omega}_{\leq \mu}\mathcal{A}_i$.

By the trivialization of $E|_F$ fixed in §1.1 we identify $(E|_F \otimes \mathcal{A}, I)_x$, $x \in F$, with $(\mathcal{A}^{4d}, I_0 \oplus (-I_0))$ as a Hilbert $\mathcal{A}$-module with a skew-hermitian structure, and $(E \otimes \hat{\Omega}_{\leq \mu}\mathcal{A}_i)_x$, $x \in F$, with $(\hat{\Omega}_{\leq \mu}\mathcal{A}_i)^{4d}$ as a right $\hat{\Omega}_{\leq \mu}\mathcal{A}_i$-module with an $\hat{\Omega}_{\leq \mu}\mathcal{A}_i$-valued non-degenerated product.

Recall that the Hilbert $\mathcal{A}$-module $L^2(M, E \otimes \mathcal{A})$ can be defined as the completion of $C_c^\infty(M, E \otimes \mathcal{A})$ with respect to the norm induced by the $\mathcal{A}$-valued scalar product

$$\langle f, g \rangle := \int_M \langle f(x), g(x) \rangle \, dx \ .$$

(For fixing notation a short introduction to Hilbert $\mathcal{A}$-modules can be found in §5.1.1.) By Prop. 5.1.18 any orthonormal basis of the Hilbert space $L^2(M, E)$ is an orthonormal basis of $L^2(M, E \otimes \mathcal{A})$ and $L^2(M, E \otimes \mathcal{A})$ is isomorphic to $l^2(\mathcal{A})$.

We introduce the Schwartz space of sections of $E \otimes \hat{\Omega}_{\leq \mu}\mathcal{A}_i$:
For $k \in \mathbb{Z}/6$ define the Schwartz space

$$\mathcal{S}(Z_k, (\hat{\Omega}_{\leq \mu}\mathcal{A}_i)^{4d}) := \mathcal{S}(\mathbb{R}) \otimes_\pi C^\infty([0,1], (\hat{\Omega}_{\leq \mu}\mathcal{A}_i)^{4d}) \ .$$

For $r \geq 0$, $0 \leq b \leq \frac{1}{4}$ choose a partition of unity $\{\phi_k\}_{k \in J}$ subordinate to the covering $\mathcal{U}(r, b)$.

The embedding $F_k(r, b) \hookrightarrow Z_k$, $k \in \mathbb{Z}/6$, and the trivialization of $E|_F$ induce a map

$$C^\infty(M, E \otimes \hat{\Omega}_{\leq \mu}\mathcal{A}_i) \to C^\infty(Z_k, (\hat{\Omega}_{\leq \mu}\mathcal{A}_i)^{4d}), \ f \mapsto \phi_k f \ .$$

As a vector space let $\mathcal{S}(M, E \otimes \hat{\Omega}_{\leq \mu}\mathcal{A}_i)$ be the largest subspace of $C^\infty(M, E \otimes \hat{\Omega}_{\leq \mu}\mathcal{A}_i)$ such that for all $k \in \mathbb{Z}/6$ the maps

$$\mathcal{S}(M, E \otimes \hat{\Omega}_{\leq \mu}\mathcal{A}_i) \to \mathcal{S}(Z_k, (\hat{\Omega}_{\leq \mu}\mathcal{A}_i)^{4d}), \ f \mapsto \phi_k f$$

are well-defined. Endowed with the seminorms of $C^\infty(M, E \otimes \hat{\Omega}_{\leq \mu} \mathcal{A}_i)$ and those of $\mathcal{S}(Z_k, (\hat{\Omega}_{\leq \mu} \mathcal{A}_i)^{4d})$ applied to $\phi_k f$ for $k \in \mathbb{Z}/6$ the space $\mathcal{S}(M, E \otimes \hat{\Omega}_{\leq \mu} \mathcal{A}_i)$ is a Fréchet space. The topology does not depend on the choice of r and b, neither on the choice of the partition of unity.

To a triple $R = (P_0, P_1, P_2)$ of Lagrangian projections of $\mathcal{A}^{2d}$ with $P_i \in M_{2d}(\mathcal{A}_\infty)$, $i = 0, 1, 2$, we associate boundary conditions on sections of $E \otimes \hat{\Omega}_{\leq \mu} \mathcal{A}_i$ in the following way:

For $k \in \mathbb{N}_0 \cup \{\infty\}$ let

$$C_R^k(M, E \otimes \hat{\Omega}_{\leq \mu} \mathcal{A}_i) := \{f \in C^k(M, E \otimes \hat{\Omega}_{\leq \mu} \mathcal{A}_i) \mid (P_i \oplus P_i) \partial\!\!\!/_E^l f(x) = \partial\!\!\!/_E^l f(x)$$
$$\text{for } x \in \partial_i M \cup \partial_{i+3} M, \ i = 0, 1, 2; \ l \in \mathbb{N}_0, l \leq k\}$$

endowed with the subspace topology.

Analogously we define $C_{R0}^k(M, E \otimes \hat{\Omega}_{\leq \mu} \mathcal{A}_i)$ and $C_{Rc}^k(M, E \otimes \hat{\Omega}_{\leq \mu} \mathcal{A}_i)$.

Furthermore let $\mathcal{S}_R(M, E \otimes \hat{\Omega}_{\leq \mu} \mathcal{A}_i)$ be the vector space $\mathcal{S}(M, E \otimes \hat{\Omega}_{\leq \mu} \mathcal{A}_i) \cap C_R^\infty(M, E \otimes \hat{\Omega}_{\leq \mu} \mathcal{A}_i)$ with the subspace topology of $\mathcal{S}(M, E \otimes \hat{\Omega}_{\leq \mu} \mathcal{A}_i)$.

Now we introduce the operator D:

Fix a triple of pairwise transverse Lagrangian projections $R = (\mathcal{P}_0, \mathcal{P}_1, \mathcal{P}_2)$ of $\mathcal{A}^{2d}$ with $\mathcal{P}_i \in M_{2d}(\mathcal{A}_\infty)$, $i = 0, 1, 2$. We define D on $L^2(M, E \otimes \mathcal{A})$ as the closure of the Dirac operator $\partial\!\!\!/_E$ with domain $C_{Rc}^\infty(M, E \otimes \mathcal{A})$ and D^+ resp. D^- as the restriction of D to the sections of $E^+ \otimes \mathcal{A}$ resp. $E^- \otimes \mathcal{A}$.

Note that D is symmetric.

2.1.2. Comparison with D_s. Recall that $P_s = \frac{1}{2} \begin{pmatrix} 1 & 1 \\ 1 & 1 \end{pmatrix} \in M_{2d}(\mathbb{C})$.

Fix a triple of pairwise transverse Lagrangian projections $(\mathcal{P}_0^s, \mathcal{P}_1^s, \mathcal{P}_2^s)$ with $\mathcal{P}_0^s = P_s$ and $\mathcal{P}_1^s, \mathcal{P}_2^s \in M_{2d}(\mathbb{C})$. Let D_s on $L^2(M, E \otimes \mathcal{A})$ be the closure of the Dirac operator $\partial\!\!\!/_E$ with domain $C_{Rc}^\infty(M, E \otimes \mathcal{A})$ for the triple $R = (\mathcal{P}_0^s, \mathcal{P}_1^s, \mathcal{P}_2^s)$.

Let $W \in C^\infty(M, \text{End}^+ E \otimes \mathcal{A}_\infty)$ be such that

- $WW^* = 1$,
- $W(x)(\mathcal{P}_i \oplus \mathcal{P}_i) W(x)^* = (\mathcal{P}_i^s \oplus \mathcal{P}_i^s)$ for all $x \in \partial_i M \cup \partial_{i+3} M$, $i = 0, 1, 2$,
- W is parallel on $M \setminus F$ and on a neighborhood of ∂M,
- for all $k \in \mathbb{Z}/6$ the restriction of W to $F_k(0, \frac{1}{4})$ depends only on the coordinate x_2^k,
- W commutes with the Clifford multiplication.

Some of these properties are not needed in this section, but are important for the proof of the index theorem.

PROPOSITION 2.1.1. (1) *We have that* $WDW^* = D_s + Wc(dW^*)$ *with* $c(dW^*)|_F := c(e_2) \partial_{e_2} W^*$ *and* $c(dW^*)|_{M \setminus F} := 0$. *In particular,* $Wc(dW^*) \in C^\infty(M, \text{End}^- E \otimes \mathcal{A}_\infty)$.

(2) *The operator D is regular and selfadjoint.*

PROOF. (1) For $R = (\mathcal{P}_0^s, \mathcal{P}_1^s, \mathcal{P}_2^s)$ and $f \in C_{cR}^\infty(M, E \otimes \mathcal{A})$ we have that $(WDW^* f)|_{M \setminus F} = (D_s f)|_{M \setminus F}$ and

$$(WDW^* f)|_F = (D_s f)|_F + W[c(e_2) \partial_{e_2}, W^*]_s (f|_F)$$
$$= (D_s f)|_F + Wc(e_2)(\partial_{e_2} W^*)(f|_F).$$

(2) The restriction of D_s to the Hilbert space $L^2(M, E)$ is selfadjoint [**BK**]. Hence $(1+D_s^2)$ has a bounded inverse on $L^2(M, E)$. It follows that the range of $(1+D_s^2)$ on $L^2(M, E \otimes \mathcal{A})$ is dense, thus D_s is regular. By an analogous argument the operators $D_s \pm i$ have dense range. From Lemma 5.1.15 it follows that D_s is selfadjoint.

By (1) the operator D is a bounded perturbation of D_s, hence D is selfadjoint and regular (see Prop. 5.1.11). □

The existence of a section W fulfilling the properties above is proved in the following lemma and proposition:

LEMMA 2.1.2. *There is a parallel unitary section $U \in C^\infty(M, \operatorname{End} E^+ \otimes \mathcal{A}_\infty)$ such that $U\mathcal{P}_0 U^* = \mathcal{P}_0^s = P_s$.*

PROOF. By the isomorphism $E^+ \otimes \mathcal{A} \cong S^+ \otimes (\mathcal{A}^+)^d \oplus S^- \otimes (\mathcal{A}^-)^d$ any $A \in M_{2d}(\mathcal{A})$ of the form $\begin{pmatrix} a & 0 \\ 0 & b \end{pmatrix}$ with $a, b \in M_d(\mathcal{A})$ defines a parallel section of $\operatorname{End} E^+ \otimes \mathcal{A}$.

By Lemma 1.4.3 there is a unitary $U \in M_{2d}(\mathcal{A}_\infty)$ of that form such that $U\mathcal{P}_0 U^* = \mathcal{P}_0^s = P_s$. □

PROPOSITION 2.1.3. *For any $0 < b < \frac{1}{4}$ there is $W \in C^\infty(M, \operatorname{End}^+ E \otimes \mathcal{A}_\infty)$ with the properties above and such that W is parallel on $\{x \in M \mid d(x, \partial M) > b\}$.*

PROOF. By the previous lemma we may assume $\mathcal{P}_0 = P_s$.

In the following we identify $\partial M \times [0, b]$ with $\{x \in M \mid d(x, \partial M) \le b\}$.

Since for $i = 1, 2$ the projection $\mathcal{P}_i$ is transverse to P_s, there are, by Lemma 1.4.4, smooth paths $W_i : [0, b] \to M_{2d}(\mathcal{A}_\infty)$ of unitaries with $[W_i, I_0] = 0$ such that $W_i(0)\mathcal{P}_i W_i^*(0) = \mathcal{P}_i^s$ and such that W_i is equal to the identity in a neighborhood of b and constant on $[0, \frac{b}{2}]$. They induce maps $\tilde{W}_i := \operatorname{id} \times W_i : (\partial_i M \cup \partial_{i+3} M) \times [0, b] \to M_{2d}(\mathcal{A}_\infty)$. The map

$$\cup_i (\tilde{W}_i \oplus \tilde{W}_i) : \partial M \times [0, b] \to M_{4d}(\mathcal{A}_\infty)$$

can be extended by 1 to a smooth section $W \in C^\infty(M, \operatorname{End}^+ E \otimes \mathcal{A}_\infty)$. By construction W has the right properties. □

2.2. The operator D_I on $[0,1]$

2.2.1. Definition and comparison with D_{I_s}. Let ∂ be the differentiation operator on $[0, 1]$.

For a pair $R = (P_0, P_1)$ of transverse Lagrangian projections with $P_j \in M_{2d}(\mathcal{A}_\infty)$, $j = 0, 1$, and $k \in \mathbb{N}_0 \cup \{\infty\}$ we define the function space

$$\begin{aligned} C_R^k([0,1], (\hat{\Omega}_{\le \mu} \mathcal{A}_i)^{2d}) \\ := \ & \{f \in C^k([0,1], (\hat{\Omega}_{\le \mu} \mathcal{A}_i)^{2d}) \mid P_x(I_0 \partial)^l f(x) = (I_0 \partial)^l f(x) \\ & \text{for } x = 0, 1; \ l \in \mathbb{N}_0, \ l \le k\} \end{aligned}$$

and endow it with the subspace topology.

Given a pair (P_0, P_1) of transverse Lagrangian projections on $\mathcal{A}^{2d}$ we write D_I for the closure of the operator $I_0 \partial$ on the Hilbert $\mathcal{A}$-module $L^2([0,1], \mathcal{A}^{2d})$ with domain $C_R^\infty([0,1], \mathcal{A}^{2d})$.

We write D_{I_s} if $(P_0, P_1) = (P_s, 1 - P_s)$.

PROPOSITION 2.2.1. *Let (P_0, P_1) be a pair of transverse Lagrangian projections with $P_0, P_1 \in M_{2d}(\mathcal{A}_\infty)$. Then for $0 < x_1 < x_2 < 1$ there is $U \in C^\infty([0,1], M_{2d}(\mathcal{A}_\infty))$ such that*

(1) $UU^* = 1$,
(2) $UI_0 = I_0 U$,
(3) $U(0)P_0 U(0)^* = P_s$,
(4) $U(1)P_1 U(1)^* = 1 - P_s$,
(5) *U is constant on $[0, x_1]$ and on $[x_2, 1]$.*

PROOF. By Lemma 1.4.3 there is a unitary $U_0 \in M_{2d}(\mathcal{A}_\infty)$ with $U_0 I_0 = I_0 U_0$ and $U_0 P_0 U_0^* = P_s$. Since $U_0 P_1 U_0^*$ is transverse to P_s, we can apply Lemma 1.4.4 to $P = U_0 P_1 U_0^*$ and $\tilde{P} = 1 - P_s$ in order to get a smooth path of unitaries $U_1 : [0,1] \to M_{2d}(\mathcal{A}_\infty)$ such that $U(t) := U_1(t) U_0$ has the right properties. □

PROPOSITION 2.2.2. *Let (P_0, P_1) be a pair of transverse Lagrangian projections of $\mathcal{A}^{2d}$ with $P_0, P_1 \in M_{2d}(\mathcal{A}_\infty)$ and let D_I be the associated operator.*

(1) *Let U be as in the previous proposition. Then*
$$UD_I U^* = D_{I_s} + UI_0(\partial U^*)$$
with $UI_0(\partial U^*) \in C^\infty([0,1], M_{2d}(\mathcal{A}_\infty))$.
(2) *The operator D_I is regular and selfadjoint.*

PROOF. (1) For $f \in C_R^\infty([0,1], \mathcal{A}_\infty^{2d})$ with $R = (P_s, 1 - P_s)$ we have:
$$\begin{aligned} UD_I U^* f &= UI_0 \partial U^* f \\ &= D_{I_s} f + UI_0(\partial U^*) f . \end{aligned}$$

(2) follows as in Prop. 2.1.1. □

2.2.2. Generalized eigenspace decomposition. We construct a decomposition of $L^2([0,1], \mathcal{A}^{2d})$ into free finitely generated $\mathcal{A}$-modules preserved by D_I. For $d = 1$ these can be understood as analogues of eigenspaces.

Assume that the boundary conditions of D_I are given by a pair (P_0, P_1) with $P_0 = P_s$. For the general case use Lemma 1.4.3.

Let $p \in M_d(\mathcal{A})$ be such that $P_1 = \frac{1}{2} \begin{pmatrix} 1 & p^* \\ p & 1 \end{pmatrix}$.

The transversality of P_0 and P_1 implies that $1 - p$ is invertible. Hence $\log p$ is defined for $\log : \mathbb{C} \setminus [0, \infty) \to \mathbb{C}$.

LEMMA 2.2.3. *Assume that $d = 1$.*
Then for $k \in \mathbb{Z}$ the function
$$f_k(x) = \frac{1}{\sqrt{2}}\left(\begin{pmatrix} 1 \\ 0 \end{pmatrix} \exp[(-\tfrac{1}{2}\log p + \pi k i)x] + \begin{pmatrix} 0 \\ 1 \end{pmatrix} \exp[(\tfrac{1}{2}\log p - \pi k i)x]\right)$$
is in $\operatorname{dom} D_I$ and $D_I f_k = \lambda_k f_k$ with $\lambda_k := -\tfrac{1}{2}i \log p - \pi k$.
The system $\{f_k\}_{k \in \mathbb{Z}}$ is an orthonormal basis of the Hilbert $\mathcal{A}$-module $L^2([0,1], \mathcal{A}^2)$.

Note that $\lambda_k f_k = f_k \lambda_k$ and $\sigma(\lambda_k) \subset\ ]-\pi k, -\pi(k-1)[$.

PROOF. Clearly the system $\{f_k\}_{k \in \mathbb{Z}}$ is orthonormal in the sense of Def. 5.1.17.

In order to prove that these functions form a basis we first show that there is an orthogonal projection onto the closure of their span. This implies that the span is orthogonally complemented. The second step will be to see that the complement

2.2. THE OPERATOR D_I ON $[0,1]$

is trivial. Then the claim follows from Prop. 5.1.18.

The system $\{e_l^\pm(x) := v_\pm e^{2\pi i l x}\}_{l\in\mathbb{Z},\pm}$ with $v_+ := (1,0)$, $v_- := (0,1)$ is an orthonormal basis of $L^2([0,1], \mathcal{A}^2)$ by Prop. 5.1.18.

We have that
$$\begin{aligned}
\langle f_k, e_l^\pm \rangle &= \tfrac{1}{\sqrt{2}} \int_0^1 \exp((\mp \tfrac{1}{2}\log p^* \mp \pi k i)x)\exp(2\pi i l x)\,dx \\
&= \tfrac{1}{\sqrt{2}}\left(\frac{(-1)^k \exp(\mp \tfrac{1}{2}\log p^*) - 1}{\mp \tfrac{1}{2}\log p^* + \pi i(2l \mp k)}\right),
\end{aligned}$$

hence $\{\langle f_k, e_l^\pm \rangle\}_{k\in\mathbb{Z}} \in l^2(\mathcal{A})$ for all $l \in \mathbb{Z}$ and for $\pm$. Thus
$$Pe_l^\pm := \sum_{k\in\mathbb{Z}} f_k \langle f_k, e_l^\pm \rangle \in L^2([0,1], \mathcal{A}^2)\ .$$

The $\mathcal{A}$-linear extension of P to the algebraic span of the functions $e_l^\pm$ has norm one. It follows that its closure is an orthogonal projection
$$P : L^2([0,1], \mathcal{A}^2) \to \overline{\mathrm{span}_\mathcal{A}\{f_k \mid k \in \mathbb{Z}\}}\ .$$

It remains to show that the kernel of P is trivial.

Let $g = (g_1, g_2) \in L^2([0,1], \mathcal{A}^2)$ with $\langle f_k, g \rangle = 0$ for all $k \in \mathbb{Z}$.

Hence for all $k \in \mathbb{Z}$

$$(*) \qquad \int_0^1 \exp((-\tfrac{1}{2}\log p^* - \pi k i)x)g_1(x) + \exp((\tfrac{1}{2}\log p^* + \pi k i)x)g_2(x)\,dx = 0\ .$$

Since $\exp(-\tfrac{1}{2}(\log p^*)\,x)g_1(x)$ is in $L^2([0,1], \mathcal{A})$, there is a unique $\{\lambda_l\}_{l\in\mathbb{Z}} \in l^2(\mathcal{A})$ such that
$$\sum_{l\in\mathbb{Z}} \lambda_l e^{2\pi i l x} = \exp(-\tfrac{1}{2}(\log p^*)\,x)g_1(x)\ .$$

Inserting this in $(*)$ and evaluating the integral for k even leads to
$$\lambda_{k/2} + \int_0^1 \exp((\tfrac{1}{2}\log p^* + \pi k i)x)g_2(x)dx = 0\ .$$

It follows that
$$\exp(\tfrac{1}{2}(\log p^*)\,x)g_2(x) = \sum_{l\in\mathbb{Z}}(-\lambda_l)e^{-2\pi i l x}\ .$$

Substituting again and evaluating $(*)$ for $k = 2\nu + 1$ with $\nu \in \mathbb{Z}$ we obtain
$$\begin{aligned}
0 &= \int_0^1 \sum_{l\in\mathbb{Z}} \lambda_l (e^{\pi i(2(l-\nu)-1)x} - e^{-\pi i(2(l-\nu)-1)x})\,dx \\
&= -\sum_{l\in\mathbb{Z}} \lambda_l \frac{4}{\pi i(2(l-\nu)-1)}\ .
\end{aligned}$$

Note that for every $l \in \mathbb{Z}$ the function
$$a_l : \mathbb{Z} \to \mathbb{C},\ \nu \mapsto \frac{2}{\pi i(2(l-\nu)-1)}$$

is in $l^2(\mathbb{C})$. We claim that $\{a_l\}_{l\in\mathbb{Z}}$ is an orthonormal basis of $l^2(\mathbb{C})$. Then $\{a_l\}_{l\in\mathbb{Z}}$ is an orthonormal basis of $l^2(\mathcal{A})$ as well, thus $\lambda_l = 0$ for all $l \in \mathbb{Z}$ and hence $g_1 = g_2 = 0$.

The Fourier transform of
$$a_0 : \nu \mapsto \frac{-2}{\pi i(2\nu + 1)}$$
is
$$h(x) = -ie^{-\pi i x}(1_{[0,1/2]}(x) - 1_{]1/2,1]}(x)) \in L^2(\mathbb{R}/\mathbb{Z}) \ .$$
Since $a_l(\nu) = a_0(\nu - l)$, the Fourier transform of a_l is $h(x)e^{-2\pi i l x}$.

By $hh^* = 1$ the system $\{h(x)e^{-2\pi i l x}\}_{l \in \mathbb{Z}}$ is an orthonormal basis of $L^2(\mathbb{R}/\mathbb{Z})$. This implies that $\{a_l\}_{l \in \mathbb{Z}}$ is an orthonormal basis of $l^2(\mathbb{C})$. □

For general d there is a decomposition of $L^2([0,1]\mathcal{A}^{2d})$ into $\mathcal{A}$-modules of rank d:

PROPOSITION 2.2.4. *For $k \in \mathbb{Z}$ let $U_k \subset L^2([0,1], \mathcal{A}^{2d})$ be the right $\mathcal{A}$-module spanned by the column vectors of*
$$\frac{1}{\sqrt{2}} \begin{pmatrix} \exp[(-\frac{1}{2}\log p + \pi k i)x] \\ \exp[(\frac{1}{2}\log p - \pi k i)x] \end{pmatrix} \in C^\infty([0,1], M_{2d \times d}(\mathcal{A}_\infty)) \ .$$
Each U_k is a free right $\mathcal{A}$-module of rank d and
$$L^2([0,1], \mathcal{A}^{2d}) = \widehat{\bigoplus_{k \in \mathbb{Z}}} U_k \ .$$
The sum is orthogonal.

Furthermore $U_k \subset \operatorname{dom} D_I$ and for $f \in U_k$
$$D_I f = \begin{pmatrix} \lambda_k & 0 \\ 0 & \lambda_k \end{pmatrix} f \in U_k$$
with $\lambda_k = -\frac{1}{2}i\log p - \pi k$.

PROOF. The projections $P_0, P_1 \in M_{2d}(\mathcal{A})$ are Lagrangian projections on $M_d(\mathcal{A})^2$ (see remark after Def. 1.4.2).

Let $\tilde{D}_I$ be the closure of $I_0 \partial$ on $L^2([0,1], M_d(\mathcal{A})^2)$ with domain $C_R^\infty([0,1], M_{2d}(\mathcal{A}))$ with $R = (P_0, P_1)$. Then $\tilde{D}_I = \oplus^d D_I$ with respect to the decomposition as right $M_d(\mathcal{A})$-modules
$$L^2([0,1], M_d(\mathcal{A})^2) = L^2([0,1], \mathcal{A}^{2d})^d$$
induced by the decomposition of a matrix into its column vectors.

By the previous proposition the Hilbert $M_d(\mathcal{A})$-module $L^2([0,1], M_d(\mathcal{A})^2)$ has an orthonormal basis $\{f_k\}_{k \in \mathbb{Z}}$ such that
$$\tilde{D}_I f_k = \begin{pmatrix} \lambda_k & 0 \\ 0 & \lambda_k \end{pmatrix} f_k \ .$$
For $k \in \mathbb{Z}$ let P_k be the orthogonal projection onto the span of f_k in $L^2([0,1], M_d(\mathcal{A})^2)$. It is diagonal with respect to the decomposition $L^2([0,1], M_d(\mathcal{A})^2) = L^2([0,1], \mathcal{A}^{2d})^d$.

Hence
$$L^2([0,1], \mathcal{A}^{2d}) = \widehat{\bigoplus_{k \in \mathbb{Z}}} P_k L^2([0,1], \mathcal{A}^{2d}) \ .$$
The assertion follows now since $P_k L^2([0,1], \mathcal{A}^{2d}) = U_k$. The module U_k is free since $\exp[(-\frac{1}{2}\log p + \pi k i)x]$ is invertible in $M_d(\mathcal{A})$ for all $x \in [0,1]$. □

COROLLARY 2.2.5. *Let $\lambda \in \mathbb{C}$. The operator $D_I - \lambda$ has a bounded inverse on $L^2([0,1], \mathcal{A}^{2d})$ if and only if $\exp(2i\lambda) \notin \sigma(p)$.*

PROOF. By the previous proposition $(D_I - \lambda)|_{U_k}$ is invertible if and only if $e^{2i\lambda} \notin \sigma(p)$. Furthermore for λ with $e^{2i\lambda} \notin \sigma(p)$ the inverse of $(D_I - \lambda)|_{U_k}$ is uniformly bounded in k. Hence the closure of $\oplus_k (D_I - \lambda)|_{U_k}$ has a bounded inverse by Cor. 5.1.20. In particular the closure of $\oplus_k D_I|_{U_k}$ is selfadjoint. Since D_I is a selfadjoint extension of $\oplus_k D_I|_{U_k}$, it follows that D_I is the closure of $\oplus_k D_I|_{U_k}$. Hence $D_I - \lambda$ has a bounded inverse if $\exp(2i\lambda) \notin \sigma(p)$. $\square$

2.3. The operator D_Z on the cylinder

Let $X = \mathbb{R}, \mathbb{R}/\mathbb{Z}$. Endow $X \times [0,1]$ with the euclidean metric and a spin structure and let (x_1, x_2) be the euclidian coordinates of $X \times [0,1]$. Let $S = S^+ \oplus S^-$ be the spinor bundle on $X \times [0,1]$ endowed with the Levi-Cività connection and a parallel metric. Then $S \otimes ((\mathcal{A}^+)^d \oplus (\mathcal{A}^-)^d)$ is Dirac bundle on $X \times [0,1]$ with the $\mathcal{A}$-valued scalar product induced by the hermitian metric on S and the standard $\mathcal{A}$-valued scalar product on $\mathcal{A}^d$. Let $\slashed{\partial}_Z$ be the associated Dirac operator.

We choose a parallel unit section s of S^+ and identify $S \otimes ((\mathcal{A}^+)^d \oplus (\mathcal{A}^-)^d)$ with the trivial bundle $(X \times [0,1]) \times ((\mathcal{A}^+)^{2d} \oplus (\mathcal{A}^-)^{2d})$ via the isomorphisms

$$S_x^+ \otimes (\mathcal{A}^+)^d \oplus S_x^- \otimes (\mathcal{A}^-)^d \to (\mathcal{A}^+)^d \oplus (\mathcal{A}^+)^d ,$$

$$(s(x) \otimes v) \oplus (ic(dx_1)s(x) \otimes w) \mapsto (v, w) ,$$

and

$$S_x^- \otimes (\mathcal{A}^+)^d \oplus S_x^+ \otimes (\mathcal{A}^-)^d \to (\mathcal{A}^-)^d \oplus (\mathcal{A}^-)^d ,$$

$$(ic(dx_1)s(x) \otimes v) \oplus (s(x) \otimes w) \mapsto (v, w) .$$

for $x \in X \times [0,1]$.

Let $I := c(dx_2)c(dx_1) = I_0 \oplus (-I_0)$. Then

$$\slashed{\partial}_Z = c(dx_1)(\partial_{x_1} + I\partial_{x_2}) .$$

Given a pair (P_0, P_1) of transverse Lagrangian projections of $\mathcal{A}^{2d}$ with $P_i \in M_{2d}(\mathcal{A}_\infty)$ we define D_Z to be the closure of $\slashed{\partial}_Z$ with domain

$$\{f \in C_c^\infty(X \times [0,1], \mathcal{A}_\infty^{4d}) \mid (P_i \oplus P_i)f(x,i) = f(x,i) \text{ for all } x \in X, \ i = 0, 1\}$$

on the Hilbert $\mathcal{A}$-module $L^2(X \times [0,1], \mathcal{A}^{4d})$.

We write $H(D_Z)$ for the Hilbert $\mathcal{A}$-module whose underlying right $\mathcal{A}$-module is $\operatorname{dom} D_Z$ and whose $\mathcal{A}$-valued scalar product is $\langle f, g \rangle_{D_Z} = \langle f, g \rangle + \langle D_Z f, D_Z g \rangle$ (see §5.1.3).

PROPOSITION 2.3.1. (1) *The operator D_Z is selfadjoint on $L^2(X \times [0,1], \mathcal{A}^{4d})$ and has a bounded inverse.*
(2) *If $X = \mathbb{R}/\mathbb{Z}$, then the inclusion $\iota : H(D_Z) \to L^2(\mathbb{R}/\mathbb{Z} \times [0,1], \mathcal{A}^{4d})$ is compact.*

PROOF. The proof is analogous to the family case [**BK**].
(1) By Lemma 1.4.3 we may assume $P_0 = P_s$. Recall from Prop. 2.2.4 that the operator D_I with boundary conditions (P_0, P_1) induces a decomposition

$$L^2([0,1], (\mathcal{A}^+)^{2d}) = \widehat{\bigoplus_{l \in \mathbb{Z}}} U_l$$

such that for each $l \in \mathbb{Z}$ there is $\lambda_l \in M_d(\mathcal{A})$ with $D_I f = \begin{pmatrix} \lambda_l & 0 \\ 0 & \lambda_l \end{pmatrix} f$ for $f \in U_l$.

We define submodules $U_{l,+} := U_l$ and $U_{l,-} := ic(dx_1)U_l$ of $L^2([0,1], \mathcal{A}^{4d})$. Then
$$\bigoplus_{l \in \mathbb{Z}}(U_{l,+} \oplus U_{l,-})L^2(X)$$
is dense in $L^2(X \times [0,1], \mathcal{A}^{4d})$.

First consider the case $X = \mathbb{R}$:

For $l \in \mathbb{Z}$ let $\partial_{l,\pm}$ be the closure of the unbounded operator $(\partial_{x_1} \pm \lambda_l)$ on $U_{l,\pm}L^2(\mathbb{R})$ with domain $U_{l,\pm}\mathcal{S}(\mathbb{R})$ and let ∂_e be the closure of $c(dx_1)(\bigoplus_{l \in \mathbb{Z}, \pm} \partial_{l,\pm})$.

The operator D_Z is an extension of ∂_e.

We claim that ∂_e has a bounded inverse on $L^2(\mathbb{R} \times [0,1], \mathcal{A}^{4d})$.

The Fourier transform on $L^2(\mathbb{R})$ induces an automorphism on $U_{l,\pm}L^2(\mathbb{R})$. Conjugation by it transforms $\partial_{l,\pm}$ into multiplication by
$$ix_1 \pm \begin{pmatrix} \lambda_l & 0 \\ 0 & \lambda_l \end{pmatrix}.$$

Since $\sigma(\lambda_l) \subset \mathbb{R}^*$, we see that the operator $\partial_{l,\pm}$ has a bounded inverse and the norm of the inverse tends to zero for $l \to \pm\infty$. By Cor. 5.1.20 it follows that the closure of $\oplus_{l,\pm}\partial_{l,\pm}$ has a bounded inverse.

Hence the operator ∂_e has a bounded inverse as well. In particular it is selfadjoint, thus $D_Z = \partial_e$. For $X = \mathbb{R}$ the assertion follows.

If $X = \mathbb{R}/\mathbb{Z}$, then the spaces $U_{l,\pm}L^2(\mathbb{R}/\mathbb{Z})$ decompose further into a direct sum
$$U_{l,\pm}L^2(\mathbb{R}/\mathbb{Z}) = \hat{\bigoplus}_{k \in \mathbb{Z}} V_{kl,\pm}$$
with $V_{kl,\pm} := e^{2\pi i k x_1}U_{l,\pm}$. Note that $V_{kl,\pm}$ is isomorphic to $\mathcal{A}^d$ as a Hilbert $\mathcal{A}$-module.

Let $\partial_{kl,\pm} \in B(V_{kl,\pm})$ be defined by
$$\partial_{kl,\pm}f := (\partial_{x_1} \pm \lambda_l)f = \begin{pmatrix} 2\pi i k \pm \lambda_l & 0 \\ 0 & 2\pi i k \pm \lambda_l \end{pmatrix} f$$
and let ∂_e be the closure of $c(dx_1)(\bigoplus_{k,l,\pm} \partial_{kl,\pm})$.

The operator D_Z is an extension of ∂_e.

Since $|(2\pi i k \pm \lambda_l)^{-1}|$ tends to zero for $k, l \to \pm\infty$, the closure of $\oplus_{kl,\pm}\partial_{kl,\pm}$ has a compact inverse by Cor. 5.1.20 and hence ∂_e has a compact inverse as well.

Now (1) follows as above.

(2) follows from the compactness of D_Z^{-1} for $X = \mathbb{R}/\mathbb{Z}$ since $\iota = D_Z^{-1}D_Z : H(D_Z) \to L^2(\mathbb{R}/\mathbb{Z} \times [0,1], \mathcal{A}^{4d})$. $\square$

2.4. The index of D^+

Recall from §2.1.1 that the boundary conditions of D were defined by a triple $R = (\mathcal{P}_0, \mathcal{P}_1, \mathcal{P}_2)$. For an open precompact subset U of M we define $H^1_{R0}(U, E \otimes \mathcal{A})$ to be the closure of $C^\infty_{Rc}(U, E \otimes \mathcal{A})$ in $H(D)$ (see §5.1.3 for the definition of $H(D)$).

Note that $H(D)$ is isomorphic to $l^2(\mathcal{A})$ as a Hilbert $\mathcal{A}$-module, since $L^2(M, E \otimes \mathcal{A})$ is isomorphic to $l^2(\mathcal{A})$ and since $L^2(M, E \otimes \mathcal{A})$ and $H(D)$ are isomorphic by Lemma 5.1.10.

LEMMA 2.4.1. *The inclusion* $\iota : H^1_{R0}(M_r, E \otimes \mathcal{A}) \to L^2(M, E \otimes \mathcal{A})$, $r \geq 0$, *is compact.*

2.4. THE INDEX OF D^+

PROOF. Let $\mathcal{V} := \{V_k\}_{k \in L}$ be an open covering of M_r such that the index set L is a finite subset of $\mathbb{N}$ with $1 \in L$ and such that:

- $V_1 = M \setminus \overline{F(r, \frac{1}{6})}$;
- for $k > 1$ there is an isometry $V_k \cong]0, \frac{1}{2}[\times [0, \frac{1}{5}[$. In particular V_k is in the flat region and has exactly one boundary component.

First we prove that the maps $\iota_k : H^1_{R0}(V_k, E \otimes \mathcal{A}) \to L^2(V_k, E \otimes \mathcal{A})$ are compact for all $k \in L$.

The compactness of ι_1 follows from the Sobolev embedding theorem ([**MF**], Lemma 3.3).

For $k \neq 1$ let $i \in \{0, 1, 2\}$ be such that $\partial V_k \subset (\partial_i M \cup \partial_{i+3} M)$ and set $P_k := \mathcal{P}_i$. Let D_k be the operator D_Z on the bundle $(\mathbb{R}/\mathbb{Z} \times [0, 1]) \times \mathcal{A}^{4d}$ with boundary conditions given by the pair $(P_k, 1 - P_k)$.

The map ι_k is compact since it factors through the map $H(D_k) \to L^2(\mathbb{R}/\mathbb{Z} \times [0, 1], \mathcal{A}^{4d})$, which is compact by Prop. 2.3.1.

Let $\{\phi_k\}_{k \in L}$ be a smooth partition of unity subordinate to the covering $\mathcal{V}$ such that $\partial_{e_2} \phi_k(x) = 0$ for all $x \in \partial M$ and $k \in L$. Multiplication with ϕ_k is a bounded map from $H^1_{R0}(M_r, E \otimes \mathcal{A})$ to $H^1_{R0}(V_k, E \otimes \mathcal{A})$, hence by $\iota = \sum_{k \in L} \iota_k \phi_k$, the inclusion ι is compact. □

Let $H(D)^+$ be the subspace of $H(D)$ of homogeneous elements of degree zero.

PROPOSITION 2.4.2. *The operator*
$$D^+ : H(D)^+ \to L^2(M, E^- \otimes \mathcal{A})$$
is a Fredholm operator.

PROOF. By constructing a parametrix for D^+ we show that D^+ is Fredholm (see Prop. 5.1.4).

The construction of the parametrix is analogous to the construction in the family case [**BK**]:

Choose a smooth partition of unity $\{\phi_k\}_{k \in J}$ subordinate to the covering $\mathcal{U}(0, \frac{1}{4})$ (defined in §1.1) and a system of smooth functions $\{\gamma_k\}_{k \in J}$ on M such that for all $k \in J$

- $\operatorname{supp} \gamma_k \subset \mathcal{U}_k$ and $\gamma_k \phi_k = \phi_k$,
- $\partial_{e_2} \gamma_k(x) = \partial_{e_2} \phi_k(x) = 0$ for all $x \in \partial M$.

For $k \in \mathbb{Z}/6$ let D_{Z_k} be the operator defined in §2.3 on $L^2(Z_k, \mathcal{A}^{4d})$ with boundary conditions given by $(\mathcal{P}_{k \bmod 3}, \mathcal{P}_{(k+1) \bmod 3})$. By Prop. 2.3.1 it is invertible. Let
$$Q_k := D_{Z_k}^{-1} : L^2(Z_k, \mathcal{A}^{4d}) \to H(D_{Z_k}) .$$

Since the symbol of D is elliptic and since $\mathcal{U}_\clubsuit$ is precompact, there is a parametrix $Q_\clubsuit : L^2(\mathcal{U}_\clubsuit, E \otimes \mathcal{A}) \to H^1_0(\mathcal{U}_\clubsuit, E \otimes \mathcal{A})$ such that $\gamma_\clubsuit(DQ_\clubsuit - 1)\phi_\clubsuit$ resp. $\gamma_\clubsuit(Q_\clubsuit D - 1)\phi_\clubsuit$ is compact on $L^2(\mathcal{U}_\clubsuit, E \otimes \mathcal{A})$ resp. $H^1_0(\mathcal{U}_\clubsuit, E \otimes \mathcal{A})$ [**MF**]. Furthermore $Q_\clubsuit$ can be chosen to be an odd operator.

We claim that
$$Q := \sum_{k \in J} \gamma_k Q_k \phi_k : L^2(M, E \otimes \mathcal{A}) \to H(D)$$
is a parametrix of D. Then it follows that $Q^- : L^2(M, E^- \otimes \mathcal{A}) \to H(D)^+$ is a parametrix of D^+.

In the following calculations the operators D_{Z_k} and the restriction of D to $\mathcal{U}_k$ are denoted by D as well. Let $\sim$ denote equality up to compact operators.

On $L^2(M, E \otimes \mathcal{A})$

$$DQ - 1 = \sum_{k \in J}[D, \gamma_k]Q_k\phi_k + \sum_{k \in J}\gamma_k DQ_k\phi_k - 1$$
$$\sim \sum_{k \in J} c(d\gamma_k)Q_k\phi_k \ .$$

Since $c(d\gamma_k)Q_k\phi_k$ is bounded from $L^2(M, E \otimes \mathcal{A})$ to $H^1_{R0}(M_r, E \otimes \mathcal{A})$ with $r > 0$ big enough, it is a compact operator on $L^2(M, E \otimes \mathcal{A})$ by the previous lemma. Hence $DQ - 1$ is compact.

On $H(D)$

$$QD - 1 = \sum_{k \in J}\gamma_k Q_k[D, \phi_k] + \sum_{k \in J}\gamma_k Q_k D\phi_k - 1$$
$$\sim \sum_{k \in J} \gamma_k Q_k c(d\phi_k) \ .$$

Here $c(d\phi_k) : H(D) \to L^2(\mathcal{U}_k, E \otimes \mathcal{A})$ is a compact operator by the previous lemma since $\operatorname{supp} c(d\phi_k)$ is compact. Moreover $\gamma_k Q_k : L^2(\mathcal{U}_k, E \otimes \mathcal{A}) \to H(D)$ is bounded, hence $QD - 1 : H(D) \to H(D)$ is compact. $\square$

LEMMA 2.4.3. *Let $P_0, P_1, P_2 : [0,1] \to M_{2d}(\mathcal{A})$ be continuous paths of Lagrangian projections and assume that $P_0(t), P_1(t), P_2(t)$ are pairwise transverse for each $t \in [0,1]$. Let $R(t) = (P_0(t), P_1(t), P_2(t))$. We define $D(t)$ to be the closure of $\partial\!\!\!/_E$ on $L^2(M, E \otimes \mathcal{A})$ with domain $C^\infty_{R(t)c}(M, E \otimes \mathcal{A})$.*

Let $\operatorname{ind} D(t)^+$ be the index of the Fredholm operator $D(t)^+ : H(D(t))^+ \to L^2(M, E \otimes \mathcal{A})$.

Then
$$\operatorname{ind} D(0)^+ = \operatorname{ind} D(1)^+ \in K_0(\mathcal{A}) \ .$$

PROOF. There is a continuous path of unitaries $[0,1] \to C^\infty(M, \operatorname{End} E^+ \otimes \mathcal{A})$, $t \mapsto W_t$, such that $W_t D(t)^+ W_t^* = D_s^+ + W_t c(dW_t^*)$. The family
$$D_s^+ + W_t c(dW_t^*) : H(D_s)^+ \to L^2(M, E^- \otimes \mathcal{A})$$
is a continuous path of Fredholm operators, thus $\operatorname{ind} W_0 D(0)^+ W_0^* = \operatorname{ind} W_1 D(1)^+ W_1^*$ by Prop. 5.1.7. $\square$

PROPOSITION 2.4.4. *The index of $D^+ : H(D)^+ \to L^2(M, E^- \otimes \mathcal{A})$ is*
$$\operatorname{ind} D^+ = \tau(\mathcal{P}_0, \mathcal{P}_1, \mathcal{P}_2) \in K_0(\mathcal{A}) \ .$$

PROOF. The argument is analogous to the one in [**BK**].

By Lemma 2.1.2 we may assume $\mathcal{P}_0 = P_s$.

Let $a_j := i\frac{p_j+1}{p_j-1} \in M_{2d}(\mathcal{A})$. Then $\mathcal{P}_j = P(a_j)$, $j = 1, 2$, in the notation of §1.4.2 .

Let $p^+ := 1_{\{x>0\}}(a_1 - a_2)$.

From Lemma 1.4.8 it follows that
$$\tau(\mathcal{P}_0, \mathcal{P}_1, \mathcal{P}_2) = [p^+] - [1 - p^+] \in K_0(\mathcal{A})$$

and that there are continuous paths $P_1, P_2 : [0,2] \to M_{2d}(\mathcal{A})$ of Lagrangian projections with $P_j(0) = \mathcal{P}_j$, $j = 1, 2$, with $P_1(2) = P(2p^+ - 1)$ and $P_2(2) = P(1 - 2p^+)$ and such that $P_s, P_1(t), P_2(t)$ are pairwise transverse for all $t \in [0, 2]$.

For $t \in [0, 2]$ let $D(t)$ be the Dirac operator on $L^2(M, E \otimes \mathcal{A})$ whose boundary conditions are given by the triple $(P_s, P_1(t), P_2(t))$. The previous lemma implies that $\operatorname{ind} D(0)^+ = \operatorname{ind} D(2)^+$.

We show that the index of $D(2)^+$ equals $[p^+] - [1 - p^+]$:

Let $Q_0 = \frac{1}{2}\begin{pmatrix} 1 & 1 \\ 1 & 1 \end{pmatrix} \in M_2(\mathbb{C})$, $Q_1 = \frac{1}{2}\begin{pmatrix} 1 & -i \\ i & 1 \end{pmatrix}$ and $Q_2 = \frac{1}{2}\begin{pmatrix} 1 & i \\ -i & 1 \end{pmatrix}$.

Then
$$P_1(2) = (Q_1 \otimes p^+) \oplus (Q_2 \otimes (1 - p^+))$$
and
$$P_2(2) = (Q_2 \otimes p^+) \oplus (Q_1 \otimes (1 - p^+))$$
with respect to the decomposition
$$E^+ \otimes \mathcal{A} = (S \otimes p^+ \mathcal{A}^n) \oplus (S \otimes (1 - p^+) \mathcal{A}^n) .$$

The Dirac operator respects the decomposition. By [**BK**] the Dirac operator associated to the bundle $S \otimes (\mathbb{C}^+ \oplus \mathbb{C}^-)$ has index 1 if the boundary conditions are given by the triple (Q_0, Q_1, Q_2), and index -1 if they are are given by (Q_0, Q_2, Q_1). It follows that
$$\operatorname{ind} D(2)^+ = [p^+] - [1 - p^+] .$$

$\square$

2.5. A perturbation with closed range

Imitating the construction in [**BGV**], §9.5, we define a perturbation of D by a compact operator in order to obtain an operator with closed range. Then we can express the index of D^+ in terms of the kernel and cokernel of the perturbed operator.

Choose an orthonormal basis $\{\psi_i\}_{i \in \mathbb{N}} \subset L^2(M, E^-)$ such that $\psi_i \in C_c^\infty(M, E^-)$ and $\operatorname{supp} \psi_i \subset M \setminus \partial M$ for all $i \in \mathbb{N}$. By Prop. 5.1.18 this is an orthonormal basis of $L^2(M, E^- \otimes \mathcal{A})$ as well.

Since D^+ is a Fredholm operator, there is a projective $\mathcal{A}$-module $P \subset L^2(M, E^- \otimes \mathcal{A})$ and a closed $\mathcal{A}$-module $Q \subset \operatorname{Ran} D^+$ such that $P \oplus Q = L^2(M, E^- \otimes \mathcal{A})$. By Prop. 5.1.21 there is $\mathcal{N} \in \mathbb{N}$ such that $L_\mathcal{N} := \operatorname{span}_\mathcal{A}\{\psi_i \mid i = 1, \ldots, \mathcal{N}\}$ fulfills $L_\mathcal{N} + P = L^2(M, E^- \otimes \mathcal{A})$. In particular it follows that
$$L_\mathcal{N} + \operatorname{Ran} D^+ = L^2(M, E^- \otimes \mathcal{A}) .$$

Let $M' := M \cup *$ be the disjoint union of M and one isolated point. Let E'^+ be the hermitian vector bundle $E^+ \cup (* \times \mathbb{C}^\mathcal{N})$ on M', where we endow $\mathbb{C}^\mathcal{N}$ with the standard hermitian product. Let E'^- be the hermitian bundle $E^- \cup (* \times \{0\})$ and let $E' = E'^+ \oplus E'^-$. Extend D by zero to a selfadjoint odd operator D' on $L^2(M', E' \otimes \mathcal{A})$.

As D is regular, D' is regular as well.

Furthermore $D' : H(D') \to L^2(M', E' \otimes \mathcal{A})$ is a Fredholm operator and
$$\operatorname{ind} D'^+ = \operatorname{ind} D^+ + [\mathcal{A}^\mathcal{N}] .$$

We extend the standard basis of $\mathbb{C}^\mathcal{N}$ by zero to a system $\{e_k\}_{k=1,\ldots,\mathcal{N}}$ of sections of E'^+ and define a compact selfadjoint odd operator K on $L^2(M', E' \otimes \mathcal{A})$ by

$$Kf := \sum_{k=1}^{\mathcal{N}} e_k \langle \psi_k, f \rangle + \psi_k \langle e_k, f \rangle \ .$$

Set $D(\rho) := D' + \rho K$ for $\rho \in \mathbb{R}$. Then

$$\operatorname{ind} D(\rho)^+ = \operatorname{ind} D'^+ = \operatorname{ind} D^+ + [\mathcal{A}^\mathcal{N}] \ .$$

Furthermore $D(\rho)$ is regular by Prop. 5.1.11.

By construction $D(\rho)^+$ is surjective for $\rho \neq 0$. Hence by Prop. 5.1.13 and Prop. 5.1.6 its kernel is a projective submodule of $H(D')^+$ and the kernel of $D(\rho)^-$ is trivial.

Hence:

PROPOSITION 2.5.1. *For $\rho \in \mathbb{R}$ the operator $D(\rho)^+$ is a Fredholm operator with index independent of ρ. For $\rho \neq 0$*

$$\operatorname{ind} D^+ = \operatorname{ind} D(\rho)^+ - [\mathcal{A}^\mathcal{N}] = [\operatorname{Ker} D(\rho)] - [\mathcal{A}^\mathcal{N}] \ .$$

From now on we write D, E, M for D', E', M' and we extend the operator D_s by zero to the new manifold M. Furthermore we redefine the open covering $\mathcal{U}(r, b)$ from §1.1 by including the isolated point in the set $\mathcal{U}_\clubsuit$.

CHAPTER 3

Heat Semigroups and Kernels

3.1. Complex heat kernels

In this section we collect some properties of the heat kernels associated to the operators D_{I_s} and D_s, which were defined in §2.2.1 and §2.1.2, and prove some technical lemmas for further reference. The results are proved by applying standard methods of the theory of partial differential equations to the particular geometric situation. The reader might skip this section at first reading.

3.1.1. The heat kernel of $e^{-tD_{I_s}^2}$. Since D_{I_s} is selfadjoint on $L^2([0,1], \mathbb{C}^{2d})$, the operator $-D_{I_s}^2$ generates a semigroup $e^{-tD_{I_s}^2}$ on $L^2([0,1], \mathbb{C}^{2d})$. In this section we determine the corresponding family of integral kernels by using the method of images (see [**Ta**], Ch. 3, §7) and study its properties.

The space $L^2([0,1], \mathbb{C}^{2d})$ decomposes into an orthogonal sum

$$L^2([0,1], P_s\mathbb{C}^{2d}) \oplus L^2([0,1], (1-P_s)\mathbb{C}^{2d})$$

and the semigroup $e^{-tD_{I_s}^2}$ is diagonal with respect to this decomposition.

We define an embedding

$$\tilde{} : L^2([0,1], \mathbb{C}^{2d}) \to L^2(\mathbb{R}/4\mathbb{Z}, \mathbb{C}^{2d})$$

by requiring that $\tilde{}$ is a right inverse of the map

$$L^2(\mathbb{R}/4\mathbb{Z}, \mathbb{C}^{2d}) \to L^2([0,1], \mathbb{C}^{2d}), \quad f \mapsto f|_{[0,1]}$$

and that the image of $L^2([0,1], P_s\mathbb{C}^{2d})$ resp. $L^2([0,1], (1-P_s)\mathbb{C}^{2d})$ consists of functions that are even resp. odd with respect to $y = 0$ and $y = 2$ and odd resp. even with respect to $y = 1$ and $y = 3$.

Then $\tilde{}$ maps $C_R^\infty([0,1], \mathbb{C}^{2d})$ with $R = (P_s, 1-P_s)$ into $C^\infty(\mathbb{R}/4\mathbb{Z}, \mathbb{C}^{2d})$ since

$$C_R^\infty([0,1], \mathbb{C}^{2d}) = C_l^\infty([0,1], P_s\mathbb{C}^{2d}) \oplus C_r^\infty([0,1], (1-P_s)\mathbb{C}^{2d}) \ .$$

Here

$$C_l^\infty([0,1], P_s\mathbb{C}^{2d})$$
$$:= \{f \in C^\infty([0,1], P_s\mathbb{C}^{2d}) \mid (i\partial)^{2k}f(1) = 0,\ (i\partial)^{2k+1}f(0) = 0\ \forall k \in \mathbb{N}_0\}$$

and

$$C_r^\infty([0,1], (1-P_s)\mathbb{C}^{2d})$$
$$:= \{f \in C^\infty([0,1], (1-P_s)\mathbb{C}^{2d}) \mid (i\partial)^{2k}f(0) = 0,\ (i\partial)^{2k+1}f(1) = 0\ \forall k \in \mathbb{N}_0\} \ .$$

The scalar heat kernel of $e^{t\partial^2}$ on $\mathbb{R}/4\mathbb{Z}$ is

$$H(t,x,y) = (4\pi t)^{-\frac{1}{2}} \sum_{k \in \mathbb{Z}} e^{-\frac{(x-y+4k)^2}{4t}} \ .$$

For $f \in L^2([0,1], (1-P_s)\mathbb{C}^{2d})$ and $x \in [0,1]$
$$\begin{aligned}(e^{-tD_{I_s}^2}f)(x) &= (e^{t\partial^2}\tilde{f})(x) \\ &= \int_0^1 H(t,x,y)f(y)dy + \int_1^2 H(t,x,y)f(2-y)dy \\ &\quad + \int_2^3 H(t,x,y)(-f(y-2))dy + \int_3^4 H(t,x,y)(-f(4-y))dy \;.\end{aligned}$$

It follows that the action of $e^{-tD_{I_s}^2}$ on the space $L^2([0,1],(1-P_s)\mathbb{C}^{2d})$ is given by the scalar integral kernel
$$(x,y) \mapsto H(t,x,y) + H(t,x,2-y) - H(t,x,y+2) - H(t,x,4-y) \;.$$

Analogously we conclude that the action of $e^{-tD_{I_s}^2}$ restricted to $L^2([0,1],P_s\mathbb{C}^{2d})$ is given by the integral kernel
$$(x,y) \mapsto H(t,x,y) - H(t,x,2-y) - H(t,x,y+2) + H(t,x,4-y) \;.$$

This yields the integral kernel k_t of $e^{-tD_{I_s}^2}$.

In the following we write $C_R^\infty([0,1], M_{2d}(\mathbb{C}))$ for the space of functions in $C^\infty([0,1], M_{2d}(\mathbb{C}))$ with column vectors in $C_R^\infty([0,1], \mathbb{C}^{2d})$.

LEMMA 3.1.1. *The map*
$$(0,\infty) \to C^\infty([0,1], C_R^\infty([0,1], M_{2d}(\mathbb{C}))), \; t \mapsto \big(y \mapsto k_t(\cdot, y)\big) \;,$$
is smooth.

For $\phi, \psi \in C^\infty([0,1])$ with $\operatorname{supp}\phi \cap \operatorname{supp}\psi = \emptyset$ the map $t \mapsto \big(y \mapsto \phi(\cdot)k_t(\cdot,y)\psi(y)\big)$ can be extended by zero to a smooth map from $[0,\infty)$ to $C^\infty([0,1], C_R^\infty([0,1], M_{2d}(\mathbb{C})))$.

PROOF. This follows from the corresponding properties of H. □

LEMMA 3.1.2. *Let $m, n \in \mathbb{N}_0$. Then there is $C > 0$ such that for all $x, y \in [0,1]$ and all $t > 0$*
$$|\partial_x^m \partial_y^n k_t(x,y)| \leq C(1 + t^{-\frac{m+n+1}{2}})e^{-\frac{d(x,y)^2}{4t}} \;.$$

PROOF. The assertion follows from the explicit formula of H above. When estimating the derivatives we take into account that for all $m \in \mathbb{N}$ the function $(x,y,t) \mapsto \frac{(x-y)^{2m}}{t^m}e^{-\frac{(x-y)^2}{4t}}$ can be continuously extended by zero to $t = 0$. □

3.1.2. The heat kernel of $e^{-tD_s^2}$. The operator D_s is selfadjoint on the Hilbert space $L^2(M, E)$. Hence $-D_s^2$ generates a semigroup on $L^2(M, E)$. In this section we prove the existence of the integral kernel of $e^{-tD_s^2}$ and study its properties.

In parallel we study the semigroup $e^{-tD_Z^2}$ on $L^2(Z, \mathbb{C}^{4d})$ where $Z = \mathbb{R} \times [0,1]$ and D_Z is the operator defined in §2.3. Here we assume that the boundary conditions are given by a pair (P_0, P_1) with $P_0, P_1 \in M_{2d}(\mathbb{C})$. Then D_Z is selfadjoint on $L^2(Z, \mathbb{C}^{4d})$. We will compare $e^{-tD_s^2}$ on the cylindric ends with $e^{-tD_Z^2}$ with appropriate boundary conditions in order to get estimates for the integral kernel of $e^{-tD_s^2}$.

Since the proofs are standard, they are only sketched here.

Recall that a solution $u : \mathbb{R} \to \operatorname{dom} D_s$ of the initial-value problem
$$\frac{d}{dt}u(t) = iD_s u(t), \; u(0) = f$$

with $f \in C_{Rc}^\infty(M, E)$ is unique by an energy estimate. An analogous statement holds for a solution of the corresponding problem for D_Z.

LEMMA 3.1.3. *If $f \in C_{Rc}^\infty(M, E)$, then $d(x, \operatorname{supp} f) \leq |t|$ for any $x \in \operatorname{supp}(e^{itD_s} f)$. An analogous result holds for D_Z on $L^2(Z, \mathbb{C}^{4d})$.*

This property is called "finite propagation speed property".

PROOF. The proof relies on a cutting-and-pasting argument.

For $j \in \mathbb{Z}/6$ let $V_j := \{x \in M \mid d(x, \partial_j M) < \frac{1}{4}\}$ and let $W := M \setminus \partial M$. These sets define an open covering of M. We show that the finite propagation speed property holds on these sets for small times. From this we conclude that it holds on M for all times.

By an oriented isometry we identify V_j with $\{0 \leq x_2 < \frac{1}{4}\} \subset \mathbb{R}^2$. Recall that $E|_{V_j}$ was identified with the trivial bundle with fiber $\mathbb{C}^{4d}$ in §1.1.

The restriction of $\partial\!\!\!/_E$ to V_j extends to a translation invariant differential operator $\partial\!\!\!/_{\mathbb{R}^2} : C_c^\infty(\mathbb{R}^2, \mathbb{C}^{4d}) \to L^2(\mathbb{R}^2, \mathbb{C}^{4d})$. Let $D_{\mathbb{R}^2}$ be the closure of $\partial\!\!\!/_{\mathbb{R}^2}$.

We define an embedding

$$\tilde{\ } : C_{Rc}^\infty(V_j, \mathbb{C}^{4d}) \hookrightarrow C_c^\infty(\mathbb{R}^2, \mathbb{C}^{4d})$$

intertwining the operators D_s and $D_{\mathbb{R}^2}$ using the method of images similar to §3.1.1.

Recall that the triple defining the boundary conditions of D_s was denoted by $(\mathcal{P}_0^s, \mathcal{P}_1^s, \mathcal{P}_2^s)$. Let

$$C_l^\infty(V_j, \mathcal{P}_{j \bmod 3}^s \mathbb{C}^{2d})$$
$$:= \{f \in C_c^\infty(V_j, \mathcal{P}_{j \bmod 3}^s \mathbb{C}^{2d}) \mid (\partial_{e_2}^{2k+1} f)(x) = 0, \ \forall k \in \mathbb{N}_0, \ \forall x \in \partial_j M\}$$

and

$$C_r^\infty(V_j, (1 - \mathcal{P}_{j \bmod 3}^s)\mathbb{C}^{2d})$$
$$:= \{f \in C_c^\infty(V_j, (1 - \mathcal{P}_{j \bmod 3}^s)\mathbb{C}^{2d}) \mid (\partial_{e_2}^{2k} f)(x) = 0, \ \forall k \in \mathbb{N}_0, \ \forall x \in \partial_j M\}.$$

Then $C_{Rc}^\infty(V_j, E^+)$ and $C_{Rc}^\infty(V_j, E^-)$ decompose into a direct sum

$$C_l^\infty(V_j, \mathcal{P}_{j \bmod 3}^s \mathbb{C}^{2d}) \oplus C_r^\infty(V_j, (1 - \mathcal{P}_{j \bmod 3}^s)\mathbb{C}^{2d}).$$

For $f \in C_l^\infty(V_j, \mathcal{P}_{j \bmod 3}^s \mathbb{C}^{2d})$ resp. $f \in C_r^\infty(V_j, (1 - \mathcal{P}_{j \bmod 3}^s)\mathbb{C}^{2d})$ we define $\tilde{f}$ by first extending f by zero to the half plane $\{x_2 \geq 0\}$ and then reflecting such that $\tilde{f}$ is even resp. odd with respect to $\{x_2 = 0\}$.

For $D_{\mathbb{R}^2}$ the finite propagation speed property holds. Hence the assertion of the lemma holds for all $f \in C_{Rc}^\infty(V_j, E)$ with $\operatorname{supp} f \subset \{x \in M \mid d(x, \partial_j M) < \frac{3}{16}\}$ and for $|t| < \frac{1}{16}$.

For $f \in C_c^\infty(W, E)$ with $\operatorname{supp} f \subset \{x \in M \mid d(x, \partial M) > \frac{1}{16}\}$ and for $|t| < \frac{1}{16}$ the assertion holds by the standard theory of hyperbolic equations on open subsets of $\mathbb{R}^2$.

Since every $f \in C_c^\infty(M, E)$ can be written as $f = f_W + f_0 + \ldots f_5$ with $f_W \in C_c^\infty(W, E)$ and $f_j \in C_{Rc}^\infty(V_j, E)$, the assertion holds for every $f \in C_c^\infty(M, E)$ and for $|t| < \frac{1}{16}$, and by the group property of e^{itD_s} it follows for all $t \in \mathbb{R}$.

The proof for D_Z is analogous. $\square$

For $k \in \mathbb{N}_0$ let $H^k(\mathbb{C}, D_s)$ be the Hilbert space whose underlying vector space is $\operatorname{dom} D_s^k$ and whose scalar product is given by

$$\langle f, g \rangle_{H^k} := \langle (1 + D_s^2)^{\frac{k}{2}} f, (1 + D_s^2)^{\frac{k}{2}} g \rangle .$$

Define $H^k(\mathbb{C}, D_Z)$ analogously.

LEMMA 3.1.4. *Let $k \in \mathbb{N}$, $k \geq 2$.*
  (1) *There is an embedding $H^k(\mathbb{C}, D_Z) \to C^{k-2}(Z, \mathbb{C}^{4d})$.*
  (2) *There is an embedding $H^k(\mathbb{C}, D_s) \to C^{k-2}(M, E)$.*

PROOF. We sketch the proof of (2), the proof of (1) is analogous.

Let $D_{\mathbb{R}^2}$ be as in the previous proof.

For fixed $r > 0$ the constants in the Gårding inequality for the elliptic operator $(1 + D_{\mathbb{R}^2}^2)^k$ on balls $B_r(x)$, $x \in \mathbb{R}^2$, can be chosen independent of x.

For $j \in \mathbb{Z}/6$ the embedding $C^\infty_{Rc}(V_j, E) \hookrightarrow C^\infty_c(\mathbb{R}^2, \mathbb{C}^{4d})$ defined in the previous proof intertwines the Dirac operators D_s on $C^\infty_{Rc}(V_j, E)$ and $D_{\mathbb{R}^2}$ on $C^\infty_c(\mathbb{R}^2, \mathbb{C}^{4d})$. Hence the Gårding inequality for the operator $(1+D_s^2)^k$ on balls $B_{1/8}(x) \subset M$ with $x \in \partial M$ holds with constants independent of x.

For $r > 0$ fixed and small enough we can also find global constants for the Gårding inequality for $(1 + D_s^2)^k$ on balls $B_r(x) \subset M$ with $B_r(x) \cap \partial M = \emptyset$ since M is of bounded geometry.

Then the assertion follows from the Sobolev embedding theorem. □

COROLLARY 3.1.5. *The operators $e^{-tD_Z^2}$ on $L^2(Z, \mathbb{C}^{4d})$ and $e^{-tD_s^2}$ on $L^2(M, E)$ are integral operators with smooth integral kernels.*

PROOF. This follows from the previous two lemmas (see [**Ro**], Lemma 5.6). □

LEMMA 3.1.6. *Let $f : [0, \infty) \times [0, \infty) \to \mathbb{R}$ be a function and assume that for every $\varepsilon, \delta > 0$ there is $C > 0$ such that for all $r > \varepsilon$ and $t > 0$*

$$f(r, t) \leq C e^{-\frac{(r - \varepsilon/2)^2}{(4+\delta)t}} .$$

Then for all $\varepsilon, \delta > 0$ there is $C > 0$ such that for all $r > \varepsilon$ and $t > 0$

$$f(r, t) \leq C e^{-\frac{r^2}{(4+\delta)t}} .$$

PROOF. Choose $0 < a < 1$ with $\frac{1-a}{4+\delta/2} > \frac{1}{4+\delta}$ and let $m > \frac{2}{a}$.

Then there is $C > 0$ such that for all $r > \varepsilon$ and $t > 0$

$$f(r, t) \leq C e^{-\frac{(r - \varepsilon/m)^2}{(4+\delta/2)t}} .$$

It follows that

$$\begin{aligned} f(r, t) &\leq C e^{-\frac{(1-a)r^2}{(4+\delta/2)t}} e^{\frac{r}{(4+\delta/2)t}(-ar + \frac{2\varepsilon}{m})} e^{-\frac{(\varepsilon/m)^2}{(4+\delta/2)t}} \\ &\leq C e^{-\frac{r^2}{(4+\delta)t}} . \end{aligned}$$

In the last step we used the fact that $\frac{r}{(4+\delta/2)t}(-ar + \frac{2\varepsilon}{m}) < 0$ for $r > \varepsilon$. □

LEMMA 3.1.7. *Let N be closed manifold resp. let $N = M, Z$. If N is a closed manifold, let E_N be a Dirac bundle on N and let D_N be the associated Dirac operator. If $N = M$ resp. $N = Z$, then let $E_N = E$ resp. $E_N = Z \times \mathbb{C}^{4d}$ and let $D_N = D_s$ resp. $D_N = D_Z$. Let k_t be the integral kernel of $e^{-tD_N^2}$.*

For every $\varepsilon, \delta > 0$ there is $C > 0$ such that for all $t > 0$, $r > \varepsilon$ and $x \in N$

$$\int_{N \setminus B_r(x)} |k_t(x,y)|^2 \, dy \leq C e^{-\frac{r^2}{(4+\delta)t}} \, .$$

Analogous estimates hold for the partial derivatives in x and y with respect to unit vector fields on N.

PROOF. Let $k \in 2\mathbb{N}$ with $k > \frac{\dim N}{2}$.

Let $S(x,\varepsilon) := \{u \in C_c^\infty(B_\varepsilon(x), E_N) \mid \|(1+D_N^2)^{-\frac{k}{2}} u\| \leq 1\}$.

Then by the Sobolev embedding theorem resp. by Lemma 3.1.4 there is $C > 0$ such that for all $x \in N$, $t > 0$ and $r > \varepsilon$

$$\int_{N \setminus B_r(x)} |k_t(x,y)|^2 \, dy \leq C \sup_{u \in S(x, \varepsilon/2)} \|e^{-tD_N^2} u\|_{N \setminus B_r(x)}^2 \, .$$

By a standard argument using the finite propagation speed property of D_N (see the proof of [**CGT**], Prop. 1.1) it follows that

$$\int_{N \setminus B_r(x)} |k_t(x,y)|^2 \, dy \leq Ct^{-1/2} \int_{r-\varepsilon/2}^\infty |(1+(\frac{d}{ids})^2)^{k/2} e^{-s^2/4t}| \, ds$$

$$= C \int_{\frac{r-\varepsilon/2}{\sqrt{t}}}^\infty |(1+t^{-1}(\frac{d}{ids'})^2)^{k/2} e^{-s'^2/4}| \, ds' \, .$$

Thus there is $l \in \mathbb{N}$ such that

$$\int_{N \setminus B_r(x)} |k_t(x,y)|^2 \, dy \leq C(1+t^{-l}) \int_{\frac{r-\varepsilon/2}{\sqrt{t}}}^\infty (1+s'^l) e^{-s'^2/4} \, ds'$$

$$\leq C(1+t^{-l}) e^{-\frac{(r-\varepsilon/2)^2}{(4+\delta/2)t}}$$

$$\leq C e^{-\frac{(r-\varepsilon/2)^2}{(4+\delta)t}} \, .$$

Then the assertion follows by applying the previous lemma to

$$f(r,t) := \sup_{x \in N} \int_{N \setminus B_r(x)} |k_t(x,y)|^2 \, dy \, .$$

For the derivatives the argument is similar. $\square$

LEMMA 3.1.8. *Let k_t be as in the previous lemma.*

For any $\varepsilon, \delta > 0$ there is $C < \infty$ such that for all $x, y \in N$ with $d(x,y) > \varepsilon$ and all $t > 0$

$$|k_t(x,y)| \leq C e^{-\frac{d(x,y)^2}{(4+\delta)t}} \, .$$

Analogous estimates hold for the partial derivatives in x and y with respect to unit vector fields on N.

PROOF. Let $S(y,\varepsilon)$ and $k \in 2\mathbb{N}$ be as in the proof of the previous lemma. By the Sobolev embedding theorem resp. Lemma 3.1.4 there is $C > 0$ such that for all $r > \varepsilon$, $t > 0$ and all $x,y \in N$ with $d(x,y) \geq r$

$$|k_t(x,y)| \leq C \sup_{u \in S(y, \varepsilon/4)} \|(1+D_N^2)^{\frac{k}{2}} e^{-tD_N^2} u\|_{N \setminus B_{r-\varepsilon/4}(y)}^2 \, .$$

As in the proof of the previous lemma this implies

$$|k_t(x,y)| \leq Ce^{-\frac{(r-\varepsilon/2)^2}{(4+\delta)t}} .$$

Then the assertion follows from Lemma 3.1.6 with $f(r,t) := \sup_{x,y \in N:\ d(x,y)=r} |k_t(x,y)|$.
For the derivatives the argument is similar. □

For the next lemma assume that $U \subset M$ is an open set for which one of the following properties holds:
  (1) U is precompact and $\overline{U} \cap \partial M = \emptyset$,
  (2) there is $k \in \mathbb{Z}/6$ such that $U \subset F_k(0, \frac{1}{4})$.

In the first case there is a closed manifold N and a Dirac bundle E_N on N such that there is a Dirac bundle isomorphism $E|_U \to E_N$ whose base map is an isometry. We identify $E|_U$ with its image in E_N. Then D_s coincides with D_N on U.

In the second case U is a subset of Z_k by §1.1. Let D_{Z_k} be the operator D_Z on $Z_k \times \mathbb{C}^{4d}$ with boundary conditions given by $(\mathcal{P}_{k \bmod 3}, \mathcal{P}_{(k+1) \bmod 3})$. Then D_s coincides with D_{Z_k} on U.

LEMMA 3.1.9. *Let U be as in (1) resp. (2). Let k_t be the integral kernel of the heat semigroup of D_s on M and let k'_t be the integral kernel of the heat semigroup of D_N resp. D_{Z_k}.*

For every $T > 0$ and $\varepsilon, \delta > 0$ there is $C > 0$ such that for all $0 < t < T$, $r > \varepsilon$ and $x, y \in U$ with $B_r(x), B_r(y) \subset U$

$$|k_t(x,y) - k'_t(x,y)| \leq Ce^{-\frac{r^2}{(4+\delta)t}} .$$

Analogous estimates hold for the partial derivatives with respect to unit vector fields on U.

PROOF. The notation is as in the proof of Lemma 3.1.7.
The estimate follows from

$$\begin{aligned}
|k_t(x,y) - k'_t(x,y)| &= \sup_{\phi \in S(x,\varepsilon)} \sup_{\psi \in S(y,\varepsilon/2)} |\langle \phi, e^{-tD_s^2}\psi\rangle - \langle \phi, e^{-tD_N^2}\psi\rangle| \\
&\leq Ct^{-\frac{1}{2}} \sup_{\phi \in S(x,\varepsilon)} \sup_{\psi \in S(y,\varepsilon/2)} |\langle \phi, \int_{\mathbb{R}} e^{-s^2/4t}(e^{isD_s} - e^{isD_N})\psi\rangle| \\
&\leq Ct^{-\frac{1}{2}} \int_{r-\varepsilon/2}^{\infty} |(1 + (\frac{d}{ids})^2)^k e^{-s^2/4t}| .
\end{aligned}$$

Here we used that $(e^{isD_s} - e^{isD_N})\psi = 0$ for $|s| \leq r - \varepsilon/2$ by the finite propagation speed property (Lemma 3.1.3) and the uniqueness of solutions of hyperbolic equations. □

3.2. The heat semigroup on closed manifolds

Let $\mathcal{B}$ be a Banach algebra with unit.

Let N be a closed manifold of dimension n. Let E_N be a Dirac bundle on N and let D_N be the associated Dirac operator. For simplicity (we will only need this case) assume that E_N is trivial as a vector bundle.

3.2. THE HEAT SEMIGROUP ON CLOSED MANIFOLDS

By Cor. 5.2.4 the associated heat kernel defines a family of bounded operators on $L^2(N, E_N \otimes \mathcal{B})$. The operators are smoothing, thus they restrict to a family of bounded operators on $C^m(N, E_N \otimes \mathcal{B})$ for any $m \in \mathbb{N}_0$. In order to show that the family extends to a holomorphic semigroup we have to study its behavior for small times.

By Lemma 5.2.10 we can define $-D_N^2$ as a closed operator on $L^2(N, E_N \otimes \mathcal{B})$ by requiring that $C^\infty(N, E_N \otimes \mathcal{B})$ is a core of $-D_N^2$.

For $t \to 0$ the heat kernel $k_t \in C^\infty(N \times N, E_N \boxtimes E_N)$ can be estimated as follows:

LEMMA 3.2.1. *Let $\varepsilon > 0$ be smaller than the injectivity radius of N and let $\chi : [0, \infty) \to [0, 1]$ be a smooth monotonously decreasing function such that $\chi(r) = 1$ for $r \leq \varepsilon/2$ and $\chi(r) = 0$ for $r \geq \varepsilon$.*
Let A be a differential operator of order m on $C^\infty(N, E_N)$.
Then there is $C > 0$ such that for all $x, y \in N$ and for all $t > 0$

$$|A_x k_t(x,y)| \leq C + Ct^{-(n+m)/2} e^{-d(x,y)^2/4t} \chi(d(x,y)) \sum_{i=0}^{m} d(x,y)^i t^{-\frac{i}{2}} .$$

PROOF. This follows from [**BGV**], Prop. 2.46, and its proof. □

PROPOSITION 3.2.2. *Let A be a differential operator of order m on $C^\infty(N, E_N \otimes \mathcal{B})$. Then there is $C > 0$ such that the action of the integral kernel $A_x k_t(x,y)$ on $L^2(N, E_N \otimes \mathcal{B})$ is bounded by $C(1 + t^{-m/2})$ for all $t > 0$.*

PROOF. Choose a finite open covering $\{U_\nu\}_{\nu \in I}$ of N of normal coordinate patches and assume that for every $x, y \in U_\nu$ the shortest geodesic connecting x and y is in U_ν.

Then there are $c_1, c_2 > 0$ such that for all $\nu \in I$ and all $x, y \in U_\nu$

$$c_1 |x - y|_\nu \leq d(x,y) \leq c_2 |x - y|_\nu ,$$

where $|\cdot|_\nu$ denotes the euclidian distance on U_ν defined by the coordinates.

Let $\{\phi_\nu\}_{\nu \in I}$ be a partition of unity subordinate to the covering $\{U_\nu\}_{\nu \in I}$.

Let $\varepsilon > 0$ be smaller than the injectivity radius of N and such that $\{x \in N \mid d(x, \operatorname{supp} \phi_\nu) \leq \varepsilon\} \subset U_\nu$ for every $\nu \in I$. Let χ be as in the previous lemma. Then

$$\phi_\nu(x) \chi(d(x,y)) \leq \phi_\nu(x) \chi(c_1 |x - y|_\nu) 1_{U_\nu}(y) .$$

By the previous lemma there is $C > 0$ such that for all $x, y \in N$ and $t > 0$ the term $|A_x k_t(x,y)|$ is bounded by

$$C + Ct^{-(n+m)/2} \sum_{\nu \in J} \phi_\nu(x) \left(e^{-c_1^2 |x-y|_\nu^2 / 4t} \chi(c_1 |x - y|_\nu) \sum_{i=0}^{m} c_2^i |x - y|_\nu^i t^{-\frac{i}{2}} \right) 1_{U_\nu}(y) .$$

The ν-th term of the outer sum is supported on $U_\nu \times U_\nu$. In the coordinates of U_ν it is of the form $\phi_\nu(x) f_t(x - y) 1_{U_\nu}(y)$ with $f_t \in L^1(\mathbb{R}^n)$, and there is $C > 0$ such that

$$\|f_t\|_{L^1} \leq Ct^{n/2}$$

for all $t > 0$.

The assertion follows now from Prop. 5.2.3 and Cor. 5.2.5. □

PROPOSITION 3.2.3. (1) *The family of integral kernels $k_t(x,y)$ defines a bounded strongly continuous semigroup on $L^2(N, E_N \otimes \mathcal{B})$, which extends to a bounded holomorphic semigroup. Its generator is $-D_N^2$.*
(2) *The family of integral kernels $k_t(x,y)$ defines a bounded strongly continuous semigroup on $C^m(N, E_N \otimes \mathcal{B})$ for every $m \in \mathbb{N}_0$.*

PROOF. (1) By the previous proposition the action of the integral kernel $k_t(x,y)$ on $L^2(N, E_N \otimes \mathcal{B})$ is uniformly bounded for $t > 0$. On $L^2(N, E_N) \odot \mathcal{B}$ it converges strongly to the identity. Thus $k_t(x,y)$ induces a bounded strongly continuous semigroup on $L^2(N, E_N \otimes \mathcal{B})$.

On $C^\infty(N, E_N \otimes \mathcal{B})$ the action of the generator coincides with the action of $-D_N^2$. Since $C^\infty(N, E_N \otimes \mathcal{B})$ is invariant under the semigroup and dense in $L^2(N, E_N \otimes \mathcal{B})$, it is a core for the generator. Hence the generator is $-D_N^2$.

By the previous proposition there is $C > 0$ such that on $L^2(N, E_N \otimes \mathcal{B})$ for all $0 < t < 1$
$$\|D_N^2 e^{-tD_N^2}\| < Ct^{-1}.$$
Since $\operatorname{Ran} e^{-tD_N^2} \subset C^\infty(N, E_N \otimes \mathcal{B}) \subset \operatorname{dom} D_N^2$ for $t > 0$, it follows, by Prop. 5.4.3, that $e^{-tD_N^2}$ extends to a holomorphic semigroup.

The integral kernel of $D_N^2 e^{-tD_N^2}$ is exponentially decaying in the supremum norm for $t \to \infty$, hence $D_N^2 e^{-tD_N^2}$ is exponentially decaying as an operator on $L^2(N, E_N \otimes \mathcal{B})$.

By Prop. 5.4.3 this shows that the holomorphic extension is bounded.

(2) follows from the fact that $k_t(x,y)$ defines a strongly continuous bounded semigroup on $C^m(N, E_N)$ by [**BGV**], Th. 2.30, and that $C^m(N, E_N \otimes \mathcal{B}) \cong C^m(N, E_N) \otimes_\varepsilon \mathcal{B}$. □

It can be deduced from the asymptotic expansion of the heat kernel ([**BGV**], Th. 2.30) that $e^{-tD_N^2}$ extends even to a holomorphic semigroup on $C^m(N, E_N \otimes \mathcal{B})$ – we do not need this fact in the following.

3.3. The heat semigroup on $[0,1]$

3.3.1. The semigroup $e^{-tD_I^2}$. Let (P_0, P_1) be a pair of transverse Lagrangian projections of $\mathcal{A}^{2d}$ with $P_0, P_1 \in M_{2d}(\mathcal{A}_\infty)$. By Lemma 5.2.10 the operator $I_0 \partial$ with domain $C_R^\infty([0,1], (\hat{\Omega}_{\le\mu} \mathcal{A}_i)^{2d})$ is closable on $L^2([0,1], (\hat{\Omega}_{\le\mu} \mathcal{A}_i)^{2d})$. In order to avoid indices its closure is denoted by D_I (compare with the operator D_I from §2.2.1).

Let $U \in M_{2d}(\mathcal{A}_\infty)$ be a unitary with $UI_0 = I_0 U$ and $UP_0 U^* = P_s$ and let $p \in M_d(\mathcal{A}_\infty)$ be such that
$$UP_1 U^* = \tfrac{1}{2} \begin{pmatrix} 1 & p^* \\ p & 1 \end{pmatrix}.$$
The unitaries U and p exist by Lemma 1.4.3.

PROPOSITION 3.3.1. *Let $\lambda \in \mathbb{C}$ with $\exp(2i\lambda) \notin \sigma(p)$.*
(1) *The operator $D_I - \lambda$ has a bounded inverse on $L^2([0,1], (\hat{\Omega}_{\le\mu} \mathcal{A}_i)^{2d})$.*
(2) *The inverse $(D_I - \lambda)^{-1}$ maps $C_R^l([0,1], (\hat{\Omega}_{\le\mu} \mathcal{A}_i)^{2d})$ isomorphically to $C_R^{l+1}([0,1], (\hat{\Omega}_{\le\mu} \mathcal{A}_i)^{2d})$ for any $l \in \mathbb{N}_0$.*
(3) *The inverse $(D_I - \lambda)^{-1}$ maps $L^2([0,1], (\hat{\Omega}_{\le\mu} \mathcal{A}_i)^{2d})$ continuously to $C([0,1], (\hat{\Omega}_{\le\mu} \mathcal{A}_i)^{2d})$.*

PROOF. The inverse of $D_I - \lambda$ is given by
$$((D_I - \lambda)^{-1} f)(x) = \int_0^x I_0 e^{-I_0 \lambda (x-y)} f(y) \, dy + \int_0^1 e^{-I_0 \lambda (x-y)} A(y) f(y) \, dy$$
with
$$A(y) = U^* \frac{i}{e^{2i\lambda} - p} \begin{pmatrix} p & e^{2i\lambda(1-y)} \\ pe^{2i\lambda y} & e^{2i\lambda} \end{pmatrix} U.$$
It is straightforward to check that this map fulfills (1), (2) and (3). □

We define D_I as an unbounded operator on $C_R^l([0,1], (\hat{\Omega}_{\leq \mu} \mathcal{A}_i)^{2d})$, $l \in \mathbb{N}_0$, by setting $\text{dom } D_I := C_R^{l+1}([0,1], (\hat{\Omega}_{\leq \mu} \mathcal{A}_i)^{2d})$. By the previous proposition D_I is a closed operator on $C_R^l([0,1], (\hat{\Omega}_{\leq \mu} \mathcal{A}_i)^{2d})$.

As before D_{I_s} denotes D_I with $R = (P_s, 1 - P_s)$.

We show that $-D_I^2$ generates a holomorphic semigroup on $L^2([0,1], (\hat{\Omega}_{\leq \mu} \mathcal{A}_i)^{2d})$ and on $C_R^l([0,1], (\hat{\Omega}_{\leq \mu} \mathcal{A}_i)^{2d})$. This will be done by first proving that $-D_{I_s}^2$ generates a holomorphic semigroup and by then applying Prop. 5.4.10.

Moreover we will deduce a norm estimate of the semigroup $e^{-tD_I^2}$ for large t from the knowledge of the resolvent set of $-D_I^2$.

We will use the same method again when we study the heat semigroup on M.

LEMMA 3.3.2. *Assume that $R = (P_s, 1 - P_s)$.*
The operator $-D_{I_s}^2$ is the generator of a bounded holomorphic semigroup $e^{-tD_{I_s}^2}$ on $L^2([0,1], (\hat{\Omega}_{\leq \mu} \mathcal{A}_i)^{2d})$ and on $C_R^l([0,1], (\hat{\Omega}_{\leq \mu} \mathcal{A}_i)^{2d})$ for any $l \in \mathbb{N}_0$.

PROOF. Let k_t be the integral kernel of $e^{-tD_{I_s}^2}$ (see §3.1.1) and let $S(t)$ be the induced integral operator.

By Lemma 3.1.2 and Prop. 5.2.3 the family $S(t)$ is uniformly bounded on $L^2([0,1], (\hat{\Omega}_{\leq \mu} \mathcal{A}_i)^{2d})$ for $t > 0$ and the family $D_{I_s}^2 S(t)$ is bounded by $C(1 + t^{-1})e^{-\omega t}$ for some $C, \omega > 0$ and all $t > 0$.

Since $S(t)$ converges strongly to the identity on $L^2([0,1], \mathbb{C}^{2d}) \odot \hat{\Omega}_{\leq \mu} \mathcal{A}_i$ for $t \to 0$ and has the semigroup property, it is a strongly continuous semigroup on $L^2([0,1], (\hat{\Omega}_{\leq \mu} \mathcal{A}_i)^{2d})$. By Prop. 5.4.3 it extends to a bounded holomorphic semigroup on $L^2([0,1], (\hat{\Omega}_{\leq \mu} \mathcal{A}_i)^{2d})$.

The kernel H from §3.1.1 defines a bounded holomorphic semigroup on $C^k(\mathbb{R}/4\mathbb{Z}, (\hat{\Omega}_{\leq \mu} \mathcal{A}_i)^{2d})$ as well by §5.2.1. This implies that $S(t)$ restricts to a bounded holomorphic semigroup on $C_R^k([0,1], (\hat{\Omega}_{\leq \mu} \mathcal{A}_i)^{2d})$. In particular it follows that $C_R^\infty([0,1], (\hat{\Omega}_{\leq \mu} \mathcal{A}_i)^{2d})$ is a core of the generator of $S(t)$. Thus the generator is $-D_{I_s}^2$. □

PROPOSITION 3.3.3. *The operator $-D_I^2$ generates a holomorphic semigroup $e^{-tD_I^2}$ on $L^2([0,1], (\hat{\Omega}_{\leq \mu} \mathcal{A}_i)^{2d})$ as well as on $C_R^k([0,1], (\hat{\Omega}_{\leq \mu} \mathcal{A}_i)^{2d})$ for all $k \in \mathbb{N}_0$, and there are $C, \omega > 0$ such that for all $t \geq 0$*
$$\|e^{-tD_I^2}\| \leq Ce^{-\omega t}$$
on $L^2([0,1], (\hat{\Omega}_{\leq \mu} \mathcal{A}_i)^{2d})$ and on $C_R^k([0,1], (\hat{\Omega}_{\leq \mu} \mathcal{A}_i)^{2d})$.

PROOF. The following arguments hold on $L^2([0,1], (\hat{\Omega}_{\leq \mu} \mathcal{A}_i)^{2d})$ as well as on $C_R^k([0,1], (\hat{\Omega}_{\leq \mu} \mathcal{A}_i)^{2d})$:

The operator $D_I - U^* D_{I_s} U$ is bounded by Prop. 2.2.2. Since $U^* D_{I_s} U$ has a bounded inverse by Prop. 3.3.1 and $-U^* D_{I_s}^2 U$ generates a bounded holomorphic semigroup by the previous lemma, we can apply Prop. 5.4.10. It follows that $-D_I^2$ generates a holomorphic semigroup.

By Prop. 3.3.1 there is $\omega > 0$ such that the spectrum of D_I^2 is a subset of $]\omega, \infty[$. Hence by Prop. 5.4.2 there is $C > 0$ with $\|e^{-tD_I^2}\| \leq Ce^{-\omega t}$. □

3.3.2. The integral kernel. In this section we prove the existence of the heat kernel for $e^{-tD_I^2}$. By cutting and pasting we construct an approximation of the semigroup $e^{-tD_I^2}$ by integral operators. Using Duhamel's principle (see Prop. 5.4.5) we prove that the error term is an integral operator as well. At the same time we obtain estimates for the integral kernel of $e^{-tD_I^2}$.

The same method will be used later in order to study the heat kernel on M (see §3.5.1) and the heat kernel associated to the superconnection (see §4.3.2).

Let $R = (P_0, P_1)$ be the boundary conditions of D_I.

Let D_{I_0} resp. D_{I_1} be defined as D_I with boundary conditions given by $(P_0, 1 - P_0)$ resp. $(1 - P_1, P_1)$.

From Lemma 1.4.3 and §3.1.1 it follows that $e^{-tD_{I_k}^2}$, $k = 0, 1$, is an integral operator for $t > 0$. Its integral kernel is denoted by $e_t^k(x, y)$.

Let $\phi_0 : [0, 1] \to [0, 1]$ be a smooth function with $\operatorname{supp} \phi_0 \subset [0, \frac{2}{3}[$ and $\operatorname{supp}(1 - \phi_0) \subset]\frac{1}{3}, 1]$ and let $\phi_1 := (1 - \phi_0)$. Furthermore choose smooth functions $\gamma_0, \gamma_1 : [0, 1] \to [0, 1]$ with

- $\gamma_k|_{\operatorname{supp} \phi_k} = 1$, $k = 0, 1$,
- $\operatorname{supp} \gamma_k' \cap \operatorname{supp} \phi_k = \emptyset$, $k = 0, 1$,
- $\operatorname{supp} \gamma_0 \subset [0, \frac{5}{6}]$ and $\operatorname{supp} \gamma_1 \subset [\frac{1}{6}, 1]$.

Write E_t for the integral operator corresponding to the integral kernel

$$e_t(x, y) := \gamma_0(x) e_t^0(x, y) \phi_0(y) + \gamma_1(x) e_t^1(x, y) \phi_1(y) .$$

Set $E_0 := 1$.

Then E_t is strongly continuous on $L^2([0,1], (\hat{\Omega}_{\leq \mu} \mathcal{A}_i)^{2d})$ as well as on $C_R^l([0,1], (\hat{\Omega}_{\leq \mu} \mathcal{A}_i)^{2d})$ at any $t \geq 0$.

For $f \in C_R^\infty([0,1], (\hat{\Omega}_{\leq \mu} \mathcal{A}_i)^{2d})$ the map $[0, \infty) \to L^2([0,1], (\hat{\Omega}_{\leq \mu} \mathcal{A}_i)^{2d})$, $t \mapsto E_t f$ is even differentiable. Hence by Duhamel's principle (Prop. 5.4.5) we have for $f \in C_R^\infty([0,1], (\hat{\Omega}_{\leq \mu} \mathcal{A}_i)^{2d})$ in $L^2([0,1], (\hat{\Omega}_{\leq \mu} \mathcal{A}_i)^{2d})$:

$$(*) \qquad e^{-tD_I^2} f - E_t f = -\int_0^t e^{-sD_I^2} \left(\frac{d}{dt} + D_I^2 \right) E_{t-s} f \, ds .$$

In the following the norm on $M_{2d}(\mathcal{A}_i)$ is denoted by $|\cdot|$.

We define $C_R^k([0,1], M_{2d}(\hat{\Omega}_{\leq \mu} \mathcal{A}_i))$, $k \in \mathbb{N}_0 \cup \{\infty\}$, as the space of functions in $C^k([0,1], M_{2d}(\hat{\Omega}_{\leq \mu} \mathcal{A}_i))$ whose column vectors are in $C_R^k([0,1], (\hat{\Omega}_{\leq \mu} \mathcal{A}_i)^{2d})$. Then any bounded operator on $C_R^k([0,1], (\hat{\Omega}_{\leq \mu} \mathcal{A}_i)^{2d})$ acts as a bounded operator on $C_R^k([0,1], M_{2d}(\hat{\Omega}_{\leq \mu} \mathcal{A}_i))$ in an obvious way.

PROPOSITION 3.3.4. *For $t > 0$ the operator $e^{-tD_I^2}$ is an integral operator. Let k_t be its integral kernel.*

(1) *The map*

$$(0, \infty) \to C^\infty([0,1], C_R^\infty([0,1], M_{2d}(\mathcal{A}_\infty))), \ t \mapsto (y \mapsto k_t(\cdot, y))$$

is well-defined and smooth.

(2) $k_t(x,y) = k_t(y,x)^*$.
(3) For every $m, n \in \mathbb{N}_0$ and every $\delta > 0$ there is $C > 0$ such that

$$|\partial_x^m \partial_y^n k_t(x,y) - \partial_x^m \partial_y^n e_t(x,y)| \leq Ct \sum_{k=0,1} e^{-\frac{d(y,\operatorname{supp}\gamma_k')^2}{(4+\delta)t}} 1_{\operatorname{supp}\phi_k}(y)$$

for all $t > 0$ and all $x, y \in [0,1]$.

PROOF. Let $f \in C_R^\infty([0,1], (\hat{\Omega}_{\leq \mu}\mathcal{A}_i)^{2d})$. From $(*)$ it follows that

$$e^{-tD_I^2} f - E_t f = \sum_{k=0,1} \int_0^t \int_0^1 e^{-sD_I^2}(\gamma_k' \partial + \partial \gamma_k') e_{t-s}^k(\cdot, y) \phi_k(y) f(y) \, dy ds \, .$$

By Lemma 3.1.1 we can extend the map

$$t \mapsto \left(y \mapsto (\gamma_k' \partial + \partial \gamma_k') e_t^k(\cdot, y) \phi_k(y)\right)$$

by zero to a smooth map from $[0, \infty)$ to $C^\infty([0,1], C_R^\infty([0,1], M_{2d}(\mathcal{A}_i)))$.

Since $e^{-sD_I^2}$ acts as a uniformly bounded operator on $C_R^\infty([0,1], M_{2d}(\mathcal{A}_i))$ by Lemma 3.1.1, it follows that the operator on the right hand side is an integral operator with smooth integral kernel.

Hence $e^{-tD_I^2}$ is an integral operator with smooth integral kernel and (1) holds. The selfadjointness of $e^{-tD_I^2}$ implies (2).

Since $d(\operatorname{supp}\phi_k, \operatorname{supp}\gamma_k') > \varepsilon$ for some $\varepsilon > 0$, there is $C > 0$, by Lemma 3.1.2, such that for all $x, y \in [0,1]$ and $t > 0$

$$|\partial_x^m \partial_y^n (k_t(x,y) - e_t(x,y))|$$

$$\leq C \sum_{k=0,1} \int_0^t \|e^{-sD_I^2}(\gamma_k' \partial + \partial \gamma_k') \partial_y^n (e_{t-s}^k(\cdot, y) \phi_k(y))\|_{C^m} \, ds \, 1_{\operatorname{supp}\phi_k}(y)$$

$$\leq C \sum_{k=0,1} \int_0^t e^{-\frac{d(y,\operatorname{supp}\gamma_k')^2}{(4+\delta)(t-s)}} \, ds \, 1_{\operatorname{supp}\phi_k}(y)$$

$$\leq C \sum_{k=0,1} t e^{-\frac{d(y,\operatorname{supp}\gamma_k')^2}{(4+\delta)t}} 1_{\operatorname{supp}\phi_k}(y) \, .$$

This shows statement (3). □

COROLLARY 3.3.5. *Let $k_t(x,y)$ be the integral kernel of $e^{-tD_I^2}$.*
For every $m, n \in \mathbb{N}_0$ and $\delta, \varepsilon > 0$ we find $C > 0$ such that for all $x, y \in [0,1]$ with $d(x,y) > \varepsilon$ and $t > 0$

$$|\partial_x^m \partial_y^n k_t(x,y)| \leq Ce^{-\frac{d(x,y)^2}{(4+\delta)t}} + Ct \sum_{k=0,1} e^{-\frac{d(y,\operatorname{supp}\gamma_k')^2}{(4+\delta)t}} 1_{\operatorname{supp}\phi_k}(y) \, .$$

PROOF. This follows from the previous proposition and Lemma 3.1.2. □

COROLLARY 3.3.6. *Let $k_t(x,y)$ be the integral kernel of $e^{-tD_I^2}$.*
Let ω be as in Prop. 3.3.3. For every $m, n \in \mathbb{N}_0$ there is $C > 0$ such that for any $t > 0$ and any $x, y \in [0,1]$

$$|\partial_x^m \partial_y^n k_t(x,y)| \leq C(1 + t^{-\frac{m+n+1}{2}}) e^{-\omega t} \, .$$

PROOF. There is $C > 0$ such that for all $x, y \in [0, 1]$ and all $0 < t < 1$

$$|\partial_x^m \partial_y^n k_t(x, y) - \partial_x^m \partial_y^n e_t(x, y)| \leq Ct \sum_{k=0,1} e^{-\frac{d(y, \operatorname{supp} \gamma_k')^2}{5t}} 1_{\operatorname{supp} \phi_k}(y) ,$$

hence, by Lemma 3.1.2,

$$|\partial_x^m \partial_y^n k_t(x, y)| \leq C(1 + t^{-\frac{m+n+1}{2}}) .$$

For all $t > 1$ and $y \in [0, 1]$

$$k_t(\cdot, y) = e^{-(t-1)D_I^2} k_1(\cdot, y) .$$

The assertion follows now since $(y \mapsto k_1(\cdot, y)) \in C^n([0, 1], C_R^m([0, 1], M_{2d}(\mathcal{A}_i)))$ and since by Prop. 3.3.3 the action of $e^{-(t-1)D_I^2}$ on $C_R^m([0, 1], M_{2d}(\mathcal{A}_i))$ is bounded by $Ce^{-\omega t}$ for some $C, \omega > 0$ and every $t > 1$. □

The following facts will be needed for the definition of the η-form.

LEMMA 3.3.7. *Let k_t be the integral kernel of $e^{-tD_{I_s}^2}$.*
Then for all $x, y \in [0, 1]$ and $t > 0$

$$\operatorname{tr}(D_{I_s})_x k_t(x, y) = 0.$$

PROOF. Let $S := 2P_s - 1 \in \operatorname{Gl}_{2d}(\mathbb{C})$. Then $S^2 = 1$, $SI + IS = 0$, $SP_s = P_s$ and $S(1 - P_s) = -(1 - P_s)$.
This implies $SD_{I_s} e^{-tD_{I_s}^2} + D_{I_s} e^{-tD_{I_s}^2} S = 0$. Therefore

$$S(D_{I_s})_x k_t(x, y) + (D_{I_s})_x k_t(x, y) S = 0 ,$$

hence

$$\operatorname{tr}(D_{I_s})_x k_t(x, y) = \operatorname{tr}(-S(D_{I_s})_x k_t(x, y) S) = -\operatorname{tr}(D_{I_s})_x k_t(x, y) .$$

It follows that $\operatorname{tr}(D_{I_s})_x k_t(x, y) = 0$. □

COROLLARY 3.3.8. *Let $(D_I k_t)$ be the integral kernel of $D_I e^{-tD_I^2}$. We have, uniformly on $[0, 1]$:*

$$\lim_{t \to 0} \operatorname{tr}(D_I k_t)(x, x) = 0 .$$

PROOF. By the previous lemma $\operatorname{tr}(D_I)_x e_t(x, y) = 0$ for all $x, y \in [0, 1]$. Then the assertion follows from the estimate in Prop. 3.3.4. □

3.4. The heat semigroup on the cylinder

Let $Z = \mathbb{R} \times [0, 1]$.
Let $R = (P_0, P_1)$ be a pair of pairwise transverse Lagrangian projections of $\mathcal{A}^{2d}$ with $P_0, P_1 \in M_{2d}(\mathcal{A}_\infty)$.
In this section we study the action of D_Z as an unbounded operator on $L^2(Z, (\hat{\Omega}_{\leq \mu} \mathcal{A}_i)^{4d})$. If not specified the notation is as in §2.3.

First we define the following function spaces and operators:
For $k \in \mathbb{N}_0 \cup \{\infty\}$ let

$$C_R^k(Z, (\hat{\Omega}_{\leq \mu} \mathcal{A}_i)^{4d})$$
$$:= \{f \in C^k(Z, (\hat{\Omega}_{\leq \mu} \mathcal{A}_i)^{4d}) \mid (P_i \oplus P_i)(\partial_Z^l f)(x, i) = f(x, i)$$
$$\text{for } x \in \mathbb{R}; \ i = 0, 1; \ l \in \mathbb{N}_0, \ l \leq k\} .$$

3.4. THE HEAT SEMIGROUP ON THE CYLINDER

Further suffixes, like c or $0 \ldots$, have their usual meaning.

We endow these spaces with the subspace topologies.

For a Fréchet space V we define the Schwartz space
$$\mathcal{S}(Z,V) := \mathcal{S}(\mathbb{R}) \otimes C^\infty([0,1],V) \ .$$
Moreover let $\mathcal{S}_R(Z,(\hat{\Omega}_{\leq \mu} \mathcal{A}_i)^{4d})$ be $\mathcal{S}(Z,(\hat{\Omega}_{\leq \mu} \mathcal{A}_i)^{4d}) \cap C_R^\infty(Z,(\hat{\Omega}_{\leq \mu} \mathcal{A}_i)^{4d})$ as a vector space with the topology induced by $\mathcal{S}(Z,(\hat{\Omega}_{\leq \mu} \mathcal{A}_i)^{4d})$.

Let D_Z as an unbounded operator on $L^2(Z,(\hat{\Omega}_{\leq \mu} \mathcal{A}_i)^{4d})$ be the closure of $\partial\!\!\!/_Z$ with domain $\mathcal{S}_R(Z,(\hat{\Omega}_{\leq \mu} \mathcal{A}_i)^{4d})$. The existence of the closure follows from Lemma 5.2.10.

Since at the moment it is not clear whether D_Z^2 is closed on $L^2(Z,(\hat{\Omega}_{\leq \mu} \mathcal{A}_i)^{4d})$, we define Δ as the closure of $\partial\!\!\!/_Z^2 = -\partial_{x_1}^2 - \partial_{x_2}^2$ with domain $\mathcal{S}_R(Z,(\hat{\Omega}_{\leq \mu} \mathcal{A}_i)^{4d})$.

Let $\Delta_\mathbb{R}$ be the closure of $-\partial_{x_1}^2$ with domain $\mathcal{S}_R(Z,(\hat{\Omega}_{\leq \mu} \mathcal{A}_i)^{4d})$.

Let $\tilde{D}_I$ be the closure of $I\partial_{x_2}$ as an unbounded operator on $L^2(Z,(\hat{\Omega}_{\leq \mu} \mathcal{A}_i)^{4d})$ with domain $\mathcal{S}_R(Z,(\hat{\Omega}_{\leq \mu} \mathcal{A}_i)^{4d})$.

By Prop. 3.3.1 the operator D_I has a bounded inverse on $L^2([0,1],(\hat{\Omega}_{\leq \mu} \mathcal{A}_i)^{2d})$. By Lemma 5.2.2 the space $L^2(Z,(\hat{\Omega}_{\leq \mu} \mathcal{A}_i)^{4d})$ can be identified with $L^2(\mathbb{R}, L^2([0,1],(\hat{\Omega}_{\leq \mu} \mathcal{A}_i)^{4d}))$, hence $\tilde{D}_I$ has a bounded inverse on $L^2(Z,(\hat{\Omega}_{\leq \mu} \mathcal{A}_i)^{4d})$. It follows that $-\tilde{D}_I^2$ is closed.

By an analogous argument $-\tilde{D}_I^2$ generates a bounded holomorphic semigroup on $L^2(Z,(\hat{\Omega}_{\leq \mu} \mathcal{A}_i)^{4d})$ with integral kernel $k_t^I(x_2,y_2) \oplus k_t^I(x_2,y_2)$ for $t > 0$. Here k_t^I is the integral kernel of $e^{-tD_I^2}$, which exists by Prop. 3.3.4.

Furthermore we have a natural candidate for the integral kernel of a semigroup generated by $-\Delta$, namely
$$k_t^Z(x,y) := \frac{1}{\sqrt{4\pi t}} \, e^{-\frac{(x_1-y_1)^2}{4t}} \left(k_t^I(x_2,y_2) \oplus k_t^I(x_2,y_2) \right) \ .$$

In the following let $\omega > 0$ be as in Prop. 3.3.3 such that
$$\|e^{-t\tilde{D}_I^2}\| \leq Ce^{-\omega t}$$
on $L^2(Z,(\hat{\Omega}_{\leq \mu} \mathcal{A}_i)^{4d})$ for all $t \geq 0$.

PROPOSITION 3.4.1.
(1) The integral kernel $k_t^Z(x,y)$ defines a holomorphic semigroup on $L^2(Z,(\hat{\Omega}_{\leq \mu} \mathcal{A}_i)^{4d})$ with generator $-\Delta$.
(2) For every $m \in \mathbb{N}_0$ the integral kernel $k_t^Z(x,y)$ defines a holomorphic semigroup on $C_R^m(Z,(\hat{\Omega}_{\leq \mu} \mathcal{A}_i)^{4d})$, denoted by $e^{-t\Delta}$ as well.
(3) Let A be a differential operator of order m with coefficients in $C^\infty(Z,M_{4d}(\hat{\Omega}_{\leq \mu} \mathcal{A}_i))$. Then for the operator $Ae^{-t\Delta}$ on $L^2(Z,(\hat{\Omega}_{\leq \mu} \mathcal{A}_i)^{4d})$ as well as for $Ae^{-t\Delta} : C_R^n(Z,(\hat{\Omega}_{\leq \mu} \mathcal{A}_i)^{4d}) \to C^n(Z,(\hat{\Omega}_{\leq \mu} \mathcal{A}_i)^{4d})$, $n \in \mathbb{N}_0$, we have:
There is $C > 0$ such that for all $t > 0$
$$\|Ae^{-t\Delta}\| \leq C(1 + t^{-m/2})e^{-\omega t} \ .$$

PROOF. (1) By Prop. 5.2.3 the kernel $\frac{1}{\sqrt{4\pi t}} e^{-\frac{(x_1-y_1)^2}{4t}}$ defines a uniformly bounded family of operators on $L^2(Z, (\hat{\Omega}_{\leq \mu}\mathcal{A}_i)^{4d})$. Since it converges strongly to the identity on $L^2(Z) \odot (\hat{\Omega}_{\leq \mu}\mathcal{A}_i)^{4d}$ for $t \to 0$, it is a strongly continuous semigroup. The space $\mathcal{S}_R(Z, (\hat{\Omega}_{\leq \mu}\mathcal{A}_i)^{4d})$ is invariant under the action of the semigroup and the action of the generator on that space is given by $\partial_{x_2}^2$. Hence the generator is $-\Delta_{\mathrm{IR}}$.

By checking the assumptions of Prop. 5.4.3 we show that the semigroup $e^{-t\Delta_{\mathrm{IR}}}$ extends to a holomorphic one:

The operator $(i\partial_{x_1})e^{-t\Delta_{\mathrm{IR}}}$ equals the convolution with the function
$$g(x_1) := \frac{-i}{\sqrt{4\pi t}} \left(\frac{x_1}{2t}\right) e^{-x_1^2/4t} .$$
Since there is $C > 0$ such that for $0 < t < 1$
$$\|g\|_{L^1} \leq Ct^{-1/2} ,$$
it follows that
$$\|(i\partial_{x_1})e^{-t\Delta_{\mathrm{IR}}}\| \leq Ct^{-1/2}$$
on $L^2(Z, (\hat{\Omega}_{\leq \mu}\mathcal{A}_i)^{4d})$, thus for $0 < t < 1$
$$\|\Delta_{\mathrm{IR}}e^{-t\Delta_{\mathrm{IR}}}\| \leq \|(i\partial_{x_1})e^{-(t/2)\Delta_{\mathrm{IR}}}\|^2 \leq Ct^{-1} .$$
Hence $-\Delta_{\mathrm{IR}}$ generates a holomorphic semigroup on $L^2(Z, (\hat{\Omega}_{\leq \mu}\mathcal{A}_i)^{4d})$.

Note that this estimate also holds on $C_R(Z, (\hat{\Omega}_{\leq \mu}\mathcal{A}_i)^{4d})$ showing that $e^{-t\Delta_{\mathrm{IR}}}$ is a holomorphic semigroup on $C_R(Z, (\hat{\Omega}_{\leq \mu}\mathcal{A}_i)^{4d})$ as well.

Since the semigroups $e^{-t\Delta_{\mathrm{IR}}}$ and $e^{-t\tilde{D}_I^2}$ commute with each other, their composition is a holomorphic semigroup. The space $\mathcal{S}_R(Z, (\hat{\Omega}_{\leq \mu}\mathcal{A}_i)^{4d})$ is invariant under the action of the semigroup and the generator acts on it as $\partial_{x_1}^2 + \partial_{x_2}^2$. Thus the generator is $-\Delta$.

(2) Since for $f \in C_R^n(Z, (\hat{\Omega}_{\leq \mu}\mathcal{A}_i)^{4d})$, $n \in \mathbb{N}$, we have that $(i\partial_{x_1})e^{-t\Delta}f = e^{-t\Delta}(i\partial_{x_1})f$ and $\tilde{D}_I e^{-t\Delta}f = e^{-t\Delta}\tilde{D}_I f$, the assertion can be reduced to the case $n = 0$.

From Prop. 3.3.3 it follows that the action of the integral kernel $k_t^I(x_2, y_2) \oplus k_t^I(x_2, y_2)$ on $C_R(Z, (\hat{\Omega}_{\leq \mu}\mathcal{A}_i)^{4d})$ extends to a holomorphic semigroup. Furthermore in (1) we showed that the integral kernel $\frac{1}{\sqrt{4\pi t}} e^{-\frac{(x_1-y_1)^2}{4t}}$ defines a holomorphic semigroup on $C_R(Z, (\hat{\Omega}_{\leq \mu}\mathcal{A}_i)^{4d})$. Hence the kernel $k_t^Z(x, y)$ defines a semigroup on $C_R(Z, (\hat{\Omega}_{\leq \mu}\mathcal{A}_i)^{4d})$ that extends to a holomorphic one.

(3) We can restrict to the case $n = 0$ by the argument in the proof of (2).

In the following the operator norms can be understood with respect to the action on $L^2(Z, (\hat{\Omega}_{\leq \mu}\mathcal{A}_i)^{4d})$ as well as with respect to the action on $C_R(Z, (\hat{\Omega}_{\leq \mu}\mathcal{A}_i)^{4d})$.

The differential operator A is a sum of operators $a_{hk}\tilde{D}_I^h(i\partial_{x_1})^k$ with $a_{hk} \in C^\infty(Z, M_{4d}(\hat{\Omega}_{\leq \mu}\mathcal{A}_i))$ and $h + k \leq m$. We have
$$\tilde{D}_I^h (i\partial_{x_1})^k e^{-t\Delta} = \tilde{D}_I^h e^{-t\tilde{D}_I^2} (i\partial_{x_1})^k e^{-t\Delta_{\mathrm{IR}}} .$$
By Cor. 5.4.9 there is $C > 0$ such that for $0 < t$
$$\|\tilde{D}_I^h e^{-t\tilde{D}_I^2}\| \leq Ct^{-h/2} e^{-\omega t} .$$
By the estimate in the proof of (1)
$$\|(i\partial_{x_1})^k e^{-t\Delta_{\mathrm{IR}}}\| \leq \|(i\partial_{x_1})e^{-(t/k)\Delta_{\mathrm{IR}}}\|^k \leq Ct^{-k/2}$$

for $0 < t < 1$.

Now the assertion follows by taking into account that $e^{-t\tilde{\Delta}_{\mathbb{R}}}$ is uniformly bounded. □

COROLLARY 3.4.2. *Let $\lambda \in \mathbb{C}$ with $\operatorname{Re}\lambda^2 < \omega$.*
(1) *The operator $D_Z - \lambda$ is invertible on $\mathcal{S}_R(Z, (\hat{\Omega}_{\leq\mu}\mathcal{A}_i)^{4d})$.*
(2) *The operator $D_Z - \lambda$ is invertible on $L^2(Z, (\hat{\Omega}_{\leq\mu}\mathcal{A}_i)^{4d})$.*
(3) $\Delta = D_Z^2$.

PROOF. (1) For any seminorm p of $\mathcal{S}_R(Z, (\hat{\Omega}_{\leq\mu}\mathcal{A}_i)^{4d})$ there are $C > 0$, $n \in \mathbb{N}$ and a seminorm q such that $p(e^{-t\Delta_{\mathbb{R}}}f) \leq C(1+t^n)q(f)$ and $p(e^{-t\tilde{D}_I^2}f) \leq Ce^{-\omega t}q(f)$ for all $f \in \mathcal{S}_R(Z, (\hat{\Omega}_{\leq\mu}\mathcal{A}_i)^{4d})$. Hence $e^{-t\Delta}$ restricts to a bounded operator on $\mathcal{S}_R(Z, (\hat{\Omega}_{\leq\mu}\mathcal{A}_i)^{4d})$, and the integral

$$G(\lambda) = \int_0^\infty (D_Z + \lambda)e^{-t(\Delta-\lambda^2)}f\,dt$$

defines a bounded operator on $\mathcal{S}_R(Z, (\hat{\Omega}_{\leq\mu}\mathcal{A}_i)^{4d})$ inverting $D_Z - \lambda$.

(2) The operator $G(\lambda)$ extends to a bounded operator on $L^2(Z, (\hat{\Omega}_{\leq\mu}\mathcal{A}_i)^{4d})$ since by the previous proposition there is $C > 0$ such that on $L^2(Z, (\hat{\Omega}_{\leq\mu}\mathcal{A}_i)^{4d})$

$$\|(D_Z + \lambda)e^{-t(\Delta-\lambda^2)}\| \leq C(1+t^{-\frac{1}{2}})e^{-(\omega-\operatorname{Re}\lambda^2)t}$$

for all $t > 0$.

From (1) it follows that $G(\lambda)$ inverts $D_Z - \lambda$ on $L^2(Z, (\hat{\Omega}_{\leq\mu}\mathcal{A}_i)^{4d})$.

(3) From (2) it follows that the operator D_Z^2 is closed, and from (1) that $\mathcal{S}_R(Z, (\hat{\Omega}_{\leq\mu}\mathcal{A}_i)^{4d})$ is a core for D_Z^2. Hence the closure of D_Z^2 equals Δ. □

PROPOSITION 3.4.3. *Let $\lambda \in \mathbb{C}$ with $\operatorname{Re}\lambda < \omega$.*
(1) *The operator $(D_Z^2 - \lambda)^{-1}$ maps $L^2(Z, (\hat{\Omega}_{\leq\mu}\mathcal{A}_i)^{4d})$ continuously to $C(Z, (\hat{\Omega}_{\leq\mu}\mathcal{A}_i)^{4d})$.*
(2) *Let $n \in \mathbb{N}, n \geq 2$. The operator $(D_Z^2 - \lambda)^{-n}$ maps $L^2(Z, (\hat{\Omega}_{\leq\mu}\mathcal{A}_i)^{4d})$ continuously to $C^{2n-3}(Z, (\hat{\Omega}_{\leq\mu}\mathcal{A}_i)^{4d})$.*

PROOF. (1) For $\operatorname{Re}\lambda < \omega$ we have on $L^2(Z, (\hat{\Omega}_{\leq\mu}\mathcal{A}_i)^{4d})$:

$$\begin{aligned}(D_Z^2 - \lambda)^{-1} &= \int_0^\infty e^{-t(D_Z^2 - \lambda)}\,dt \\ &= \int_0^\infty e^{\lambda t}e^{-t\Delta_{\mathbb{R}}^2}e^{-t\tilde{D}_I^2}\,dt\ .\end{aligned}$$

By Prop. 3.3.1 the operator $D_I^{-1} : L^2([0,1], (\hat{\Omega}_{\leq\mu}\mathcal{A}_i)^{2d}) \to C([0,1], (\hat{\Omega}_{\leq\mu}\mathcal{A}_i)^{2d})$ is bounded. Thus the family of operators

$$e^{-t\tilde{D}_I^2} = \tilde{D}_I^{-1}\tilde{D}_I e^{-t\tilde{D}_I^2} : L^2(Z, (\hat{\Omega}_{\leq\mu}\mathcal{A}_i)^{4d}) \to L^2(\mathbb{R}, C([0,1], (\hat{\Omega}_{\leq\mu}\mathcal{A}_i)^{4d}))$$

is bounded by $C(1+t^{-\frac{1}{2}})e^{-\omega t}$ for all $t > 0$.

Furthermore the family

$$e^{-t\Delta_{\mathbb{R}}} : L^2(\mathbb{R}, C([0,1], (\hat{\Omega}_{\leq\mu}\mathcal{A}_i)^{4d})) \to C(\mathbb{R}, C([0,1], (\hat{\Omega}_{\leq\mu}\mathcal{A}_i)^{4d}))$$

is bounded by $\sup_{x_1 \in \mathbb{R}} \|\frac{1}{\sqrt{4\pi t}}e^{-\frac{(x_1-y_1)^2}{4t}}\|_{L_{y_1}^2}$, hence by $Ct^{-1/4}$ for some $C > 0$.

Thus the integral converges as a bounded operator from $L^2(Z, (\hat{\Omega}_{\leq \mu} \mathcal{A}_i)^{4d})$ to $C(Z, (\hat{\Omega}_{\leq \mu} \mathcal{A}_i)^{4d})$.

(2) We show that for any $k \in \mathbb{N}_0$ with $k \leq 2n - 1$ the map

$$\tilde{D}_I^k (i\partial_{x_1})^{2n-1-k} (D_Z^2 - \lambda)^{-n} : \mathcal{S}_R(Z, (\hat{\Omega}_{\leq \mu} \mathcal{A}_i)^{4d}) \to L^2(Z, (\hat{\Omega}_{\leq \mu} \mathcal{A}_i)^{4d})$$

extends to a bounded operator on $L^2(Z, (\hat{\Omega}_{\leq \mu} \mathcal{A}_i)^{4d})$. Then the assertion follows from the first part.

We have for $k \in \mathbb{N}_0$ with $k \leq 2n + 1$

$$\begin{aligned}
&\tilde{D}_I^k (i\partial_{x_1})^{2n-k+1} (D_Z^2 - \lambda)^{-n-1} \\
&= \int_0^\infty t^{n-1} \tilde{D}_I^k (i\partial_{x_1})^{2n-k+1} e^{-t(D_Z^2 - \lambda)} \, dt \\
&= \int_0^\infty t^{n-1} \tilde{D}_I^k e^{-t(\tilde{D}_I^2 - \lambda)} (i\partial_{x_1})^{2n-k+1} e^{-t\Delta_\mathbb{R}} \, dt \; .
\end{aligned}$$

As bounded operators on $L^2(Z, (\hat{\Omega}_{\leq \mu} \mathcal{A}_i)^{4d})$

$$\|\tilde{D}_I^k e^{-t(D_I^2 - \lambda)}\| \leq C t^{-k/2} e^{-(\omega - \lambda)t}$$

and

$$\|(i\partial_{x_1})^{2n-k+1} e^{-t\Delta_\mathbb{R}}\| \leq C(1 + t^{-(2n-k+1)/2}) \; ,$$

hence the integral converges. □

In the following $|\cdot|$ denotes the norm on $M_{4d}(\mathcal{A}_i)$.

LEMMA 3.4.4. *For every $\varepsilon > 0$ and $\alpha, \beta \in \mathbb{N}_0^2$ there are $c, C > 0$ such that for all $x, y \in Z$ with $d(x, y) > \varepsilon$ and all $t > 0$*

$$|\partial_x^\alpha \partial_y^\beta k_t^Z(x, y)| \leq C e^{-\frac{d(x,y)^2}{ct}} \; .$$

PROOF. For $m, n \in \mathbb{N}_0$ there are $C, c > 0$ such that

$$|\partial_{x_2}^m \partial_{y_2}^n k_t^I(x_2, y_2)| \leq C e^{-\frac{(x_2 - y_2)^2}{ct}}$$

for $|x_2 - y_2| \geq \varepsilon/2$ and $t > 0$. This follows from Cor. 3.3.5 for $t < 1$ and from Cor. 3.3.6 for $t \geq 1$. For $|x_2 - y_2| \leq \varepsilon/2$ the left hand side is bounded by $C(1 + t^{-\frac{m+n+1}{2}})$ by Cor. 3.3.6. Similar estimates hold for $\frac{1}{\sqrt{4\pi t}} e^{-\frac{(x_1 - y_1)^2}{4t}}$. A combination of these estimates implies the lemma. □

In the next lemma $\mathcal{S}_R(Z, M_{4d}(\mathcal{A}_i))$ is the subspace of $\mathcal{S}(Z, M_{4d}(\mathcal{A}_i))$ of functions whose columns are in $\mathcal{S}_R(Z, \mathcal{A}_i^{4d})$. Then operators on $\mathcal{S}_R(Z, \mathcal{A}_i^{4d})$ act on $\mathcal{S}_R(Z, M_{4d}(\mathcal{A}_i))$ columnwise. The space $C_{Rc}^\infty(Z, M_{4d}(\mathcal{A}_i))$ is analogously defined.

LEMMA 3.4.5. *Let $\lambda \in \mathbb{C}$ with $\text{Re}\,\lambda < \omega$. Let $\xi_1, \xi_2 \in C^\infty(Z)$ with $\text{supp}\,\xi_1 \cap \text{supp}\,\xi_2 = \emptyset$ and assume that $\text{supp}\,\xi_2$ is compact.*

Then for any $n \in \mathbb{N}$ the operator $\xi_1 (D_Z^2 - \lambda)^{-n} \xi_2$ is an integral operator. Let κ be its integral kernel. Then $(y \mapsto \kappa(\cdot, y)) \in C_c^\infty(Z, \mathcal{S}_R(Z, M_{4d}(\mathcal{A}_i)))$ and $(x \mapsto \kappa(x, \cdot)^*) \in \mathcal{S}(Z, C_{Rc}^\infty(Z, M_{4d}(\mathcal{A}_i)))$.

In particular $\xi_1 (D^2 - \lambda)^{-n} \xi_2$ maps $L^2(Z, (\hat{\Omega}_{\leq \mu} \mathcal{A}_i)^{4d})$ continuously to $\mathcal{S}_R(Z, (\hat{\Omega}_{\leq \mu} \mathcal{A}_i)^{4d})$.

PROOF. First we prove the claim for $n = 1$.
Let $f \in C^\infty_{Rc}(Z, (\hat{\Omega}_{\leq \mu}\mathcal{A}_i)^{4d})$.
In $L^2(Z, (\hat{\Omega}_{\leq \mu}\mathcal{A}_i)^{4d})$

$$\begin{aligned}\xi_1(D_Z^2 - \lambda)^{-1}\xi_2 f &= \int_0^\infty \xi_1 e^{-t(D_Z^2 - \lambda)}\xi_2 f \, dt \\ &= \int_0^\infty \int_Z \xi_1 k_t^Z(\cdot, y) e^{\lambda t}\xi_2(y)f(y) \, dy dt \; .\end{aligned}$$

Let $\varepsilon > 0$ be such that $d(\operatorname{supp}\xi_1, \operatorname{supp}\xi_2) > \varepsilon$.

By the previous lemma there are $c, C > 0$ such that for all $x, y \in Z$ and all $t > 0$

$$\begin{aligned}|\xi_1(x)e^{\lambda t}k_t^Z(x,y)\xi_2(y)| &\leq C 1_{\{t>1\}}(t)|\xi_1(x)|e^{(\operatorname{Re}\lambda - \omega)t}e^{-\frac{(x_1-y_1)^2}{4t}}|\xi_2(y)| \\ &\quad + C 1_{\{t\leq 1\}}(t)|\xi_1(x)|e^{-\frac{d(x,y)^2}{ct}}|\xi_2(y)| \; .\end{aligned}$$

Analogous estimates hold for the partial derivatives. Hence we can interchange the order of integration. It follows that $\xi_1(D_Z^2 - \lambda)^{-1}\xi_2$ is an integral operator with integral kernel

$$\kappa(x,y) := \int_0^\infty \xi_1(x) e^{-\lambda t} k_t^Z(x,y)\xi_2(y) dt \; .$$

For $n = 1$ the other statements of the lemma also follow from the estimates.

For $n > 1$ choose a smooth compactly supported function $\psi : Z \to [0,1]$ such that $\operatorname{supp}\psi \cap \operatorname{supp}\xi_1 = \emptyset$ and $\operatorname{supp}(1 - \psi) \cap \operatorname{supp}\xi_2 = \emptyset$. Then

$$\begin{aligned}\xi_1(D_Z^2 - \lambda)^{-n}\xi_2 &= \xi_1(D_Z^2 - \lambda)^{-1}\psi(D_Z^2 - \lambda)^{-n+1}\xi_2 \\ &\quad + \xi_1(D_Z^2 - \lambda)^{-1}(1-\psi)(D_Z^2 - \lambda)^{-n+1}\xi_2 \; .\end{aligned}$$

By induction the lemma can be applied to $\xi_1(D_Z^2-\lambda)^{-1}\psi$ and $(1-\psi)(D_Z^2-\lambda)^{-n+1}\xi_2$. The statement of the lemma follows for $\xi_1(D_Z^2-\lambda)^{-n}\xi_2$ from this and the fact that by Cor. 3.4.2 the operator $(D_Z^2 - \lambda)^{-m}$ acts continuously on $\mathcal{S}_R(Z, M_{4d}(\mathcal{A}_i))$ for all $m \in \mathbb{N}$. □

3.5. The heat semigroup on M

3.5.1. Definitions.
Recall the definition of the operator $D(\rho)^2$ on the Hilbert $\mathcal{A}$-module $L^2(M, E \otimes \mathcal{A})$ in §2.1.1 and §2.5. By Lemma 5.2.10 the operator $(\partial\!\!\!/_E + \rho K)^2$ with domain $\mathcal{S}_R(M, E \otimes \hat{\Omega}_{\leq \mu}\mathcal{A}_i)$ is closable on $L^2(M, E \otimes \hat{\Omega}_{\leq \mu}\mathcal{A}_i)$. Its closure will be denoted by $D(\rho)^2$ as well in order to simplify notation.

So far the notation is misleading: It suggests that $D(\rho)^2$ is the square of some unbounded operator on $L^2(M, E \otimes \hat{\Omega}_{\leq \mu}\mathcal{A}_i)$. This is indeed the case as will become clear in §3.5.5.

Define D_s^2 as a closed operator on $L^2(M, E \otimes \hat{\Omega}_{\leq \mu}\mathcal{A}_i)$ in an analogous way.

We will often make use of cutting and pasting arguments on M. We fix the setting:

Let $0 < b_0 \leq \frac{1}{4}$ be small enough and $r_0 > 0$ large enough such that

$$\operatorname{supp} k_K \cap \big((F(r_0, b_0) \times M) \cup (M \times F(r_0, b_0))\big) = \emptyset \; ,$$

with $F(r_0, b_0)$ as in §1.1. Here k_K denotes the integral kernel of the operator K from §2.5.

Let $\mathcal{U}(r_0, b_0)$ be the open covering defined in §1.1 and modified in §2.5.

Choose a smooth partition of unity $\{\phi_k\}_{k \in J}$ subordinate to $\mathcal{U}(r_0, b_0)$ and smooth functions $\{\gamma_k\}_{k \in J}$ on M such that for all $k \in J$

- supp $\gamma_k \subset \mathcal{U}_k$,
- supp$(1 - \gamma_k) \cap$ supp $\phi_k = \emptyset$,
- the derivatives $\partial_{e_2}(\phi_k|_F)$ and $\partial_{e_2}(\gamma_k|_F)$ vanish in a neighborhood of ∂M.

Let E_N be a Dirac bundle on a compact spin manifold N that is trivial as a vector bundle and assume that there is a Dirac bundle isomorphism $E|_{\mathcal{U}_\clubsuit} \to E_N$, whose base map is an isometric embedding. Let D_N be the associated Dirac operator. We identify $\mathcal{U}_\clubsuit$ with its image in N and $E|_{\mathcal{U}_\clubsuit}$ with its image in E_N.

Since the support of k_K is in $\mathcal{U}_\clubsuit \times \mathcal{U}_\clubsuit$, the restriction of $D(\rho) = D + \rho K$ to $\mathcal{U}_\clubsuit$ extends to an operator $D_N + \rho K$ on the sections of E_N.

For $k \in \mathbb{Z}/6$ let D_{Z_k} be the operator D_Z on $L^2(Z, (\hat{\Omega}_{\leq \mu}\mathcal{A}_i)^{4d})$ from §3.4 with boundary conditions given by the pair $(\mathcal{P}_{k \bmod 3}, \mathcal{P}_{(k+1) \bmod 3})$.

3.5.2. The resolvents of $D(\rho)^2$. This section has three different aims:

Using a method of Lott ([**Lo3**], §6.1.) we investigate the resolvent set of $D(\rho)^2$ on $L^2(M, E \otimes \hat{\Omega}_{\leq \mu}\mathcal{A}_i)$.

Furthermore we prove a kind of Sobolev embedding theorem – more precisely an analogue of Lemma 3.1.4 for the operator $D(\rho)^2$ on $L^2(M, E \otimes \hat{\Omega}_{\leq \mu}\mathcal{A}_i)$.

Third we obtain more information about the kernel of $D(\rho)^2$ on $L^2(M, E \otimes \hat{\Omega}_{\leq \mu}\mathcal{A}_i)$, namely that there is a projection on it and that this projection is a Hilbert-Schmidt operator with a smooth integral kernel.

Let $\omega > 0$ be such that there is $C > 0$ with

$$\|e^{-tD_{Z_k}^2}\| \leq Ce^{-\omega t}$$

on $L^2(M, E \otimes \hat{\Omega}_{\leq \mu}\mathcal{A}_i)$ for all $t \geq 0$ and all $k \in \mathbb{Z}/6$.

Let $\nu \in \mathbb{N}$. For $\lambda \in \mathbb{C}$ with $\operatorname{Re}\lambda < \omega$ we define a parametrix of $(D(\rho)^2 - \lambda)^\nu$:

By Cor. 3.4.2 we can set $Q_k(\lambda) = (D_{Z_k}^2 - \lambda)^{-\nu}$ for $k \in \mathbb{Z}/6$.

Let $Q_\clubsuit(\lambda)$ be a local parametrix of $(D^2 - \lambda)^\nu$ on $\mathcal{U}_\clubsuit$ defined by the symbol of $(D^2 - \lambda)^\nu$ such that $\phi_\clubsuit(Q_\clubsuit(\lambda)(D^2 - \lambda)^\nu - 1)\gamma_\clubsuit$ and $\phi_\clubsuit((D^2 - \lambda)^\nu Q_\clubsuit(\lambda) - 1)\gamma_\clubsuit$ are integral operators with smooth integral kernels.

The operator

$$Q(\lambda) := \sum_{k \in J} \phi_k Q_k(\lambda) \gamma_k$$

acts as a bounded operator on the spaces $L^2(M, E \otimes \hat{\Omega}_{\leq \mu}\mathcal{A}_i)$ and $\mathcal{S}_R(M, E \otimes \hat{\Omega}_{\leq \mu}\mathcal{A}_i)$ by §5.2.6 and by Cor. 3.4.2.

LEMMA 3.5.1. *For any $\rho \in \mathbb{R}$ and $\lambda \in \mathbb{C}$ with $\operatorname{Re}\lambda < \omega$ the operator $Q(\lambda)(D(\rho)^2 - \lambda)^\nu - 1$ restricted to $\mathcal{S}_R(M, E \otimes \hat{\Omega}_{\leq \mu}\mathcal{A}_i)$ is an integral operator $\mathcal{K}$ with smooth integral kernel $\kappa \in L^2(M \times M, (E \boxtimes E^*) \otimes \mathcal{A}_i)$.*

Furthermore $(x \mapsto \kappa(x, \cdot)^*) \in \mathcal{S}(M, C_{cR}^\infty(M, E \otimes \mathcal{A}_i) \otimes E^*)$ and $(y \mapsto \kappa(\cdot, y)) \in C_c^\infty(M, \mathcal{S}_R(M, E \otimes \mathcal{A}_i) \otimes E^*)$.

In particular $\mathcal{K}$ extends to a bounded operator from $L^2(M, E \otimes \hat{\Omega}_{\leq \mu}\mathcal{A}_i)$ to $\mathcal{S}_R(M, E \otimes \hat{\Omega}_{\leq \mu}\mathcal{A}_i)$.

PROOF. The difference $(D(\rho)^2 - \lambda)^\nu - (D^2 - \lambda)^\nu$ on $\mathcal{S}_R(M, E \otimes \hat{\Omega}_{\leq \mu} \mathcal{A}_i)$ is an integral operator with smooth integral kernel whose support is contained in $\operatorname{supp} k_K$. Hence we only need to investigate $Q(\lambda)(D^2 - \lambda)^\nu - 1$.

For any $k \in J$ choose a function $\xi_k \in C_c^\infty(M)$ with values in $[0, 1]$ and such that $\operatorname{supp} \xi_k \subset \mathcal{U}_k$, $\xi_k|_{\operatorname{supp} d\gamma_k} = 1$ and $\operatorname{supp} \phi_k \cap \operatorname{supp} \xi_k = \emptyset$. Furthermore assume that $\partial_{e_2}(\xi_k|_F)$ vanishes in a neighborhood of ∂M.

By induction we have that $[\gamma_k, (D^2 - \lambda)^\nu] = \xi_k [\gamma_k, (D^2 - \lambda)^\nu]$ since

$$
\begin{aligned}
& [\gamma_k, (D^2 - \lambda)^\nu] \\
& = (D^2 - \lambda)^{\nu-1}\bigl(c(d\gamma_k)D + Dc(d\gamma_k)\bigr) + [\gamma_k, (D^2 - \lambda)^{\nu-1}](D^2 - \lambda) \\
& = \xi_k (D^2 - \lambda)^{\nu-1}\bigl(c(d\gamma_k)D + Dc(d\gamma_k)\bigr) + [\gamma_k, (D^2 - \lambda)^{\nu-1}](D^2 - \lambda) \ .
\end{aligned}
$$

In the following the operators D_{Z_k}, $k \in \mathbb{Z}/6$, are denoted by D as well. Furthermore $\sim$ means equality up to integral operators with smooth compactly supported integral kernels.

Then on $\mathcal{S}_R(M, E \otimes \hat{\Omega}_{\leq \mu} \mathcal{A}_i)$

$$
\begin{aligned}
Q(\lambda)(D^2 - \lambda)^\nu - 1 & = \sum_{k \in J} \phi_k Q_k(\lambda)[\gamma_k, (D^2 - \lambda)^\nu] + \sum_{k \in J} \phi_k Q_k(\lambda)(D^2 - \lambda)^\nu \gamma_k - 1 \\
& \sim \sum_{k \in J} \phi_k Q_k(\lambda) \xi_k [\gamma_k, (D^2 - \lambda)^\nu] \ .
\end{aligned}
$$

For all $k \in J$ the operator $\phi_k Q_k(\lambda) \xi_k$ is an integral operator whose integral kernel has the properties stated in the lemma. This holds for $k \in \mathbb{Z}/6$ by Lemma 3.4.5 and for $k = \clubsuit$ by the properties of pseudodifferential operators. Now the assertion follows. $\square$

PROPOSITION 3.5.2. *Let $\rho \in \mathbb{R}$. Let $\lambda \in \mathbb{C}$ with $\operatorname{Re} \lambda < \omega$ such that $D(\rho)^2 - \lambda$ has a bounded inverse on the Hilbert $\mathcal{A}$-module $L^2(M, E \otimes \mathcal{A})$.*

Then $D(\rho)^2 - \lambda$ has a bounded inverse on $L^2(M, E \otimes \hat{\Omega}_{\leq \mu} \mathcal{A}_i)$.

The inverse $(D(\rho)^2 - \lambda)^{-1}$ acts as a bounded operator on the space $\mathcal{S}_R(M, E \otimes \hat{\Omega}_{\leq \mu} \mathcal{A}_i)$.

PROOF. Let $Q(\lambda)$ and $\mathcal{K}$ be as in the previous lemma such that $Q(\lambda)(D(\rho)^2 - \lambda)f = (1 - \mathcal{K})f$ for $f \in \operatorname{dom} D(\rho)^2$.

We want to apply Prop. 5.3.1. Since in general $1 - \mathcal{K}$ is not invertible on $L^2(M, E \otimes \mathcal{A})$, we modify the parametrix:

Choose an integral kernel $s \in C_c^\infty(M \times M, (E \boxtimes E^*) \otimes \mathcal{A}_i)$ vanishing near $(\partial M \times M) \cup (M \times \partial M)$ such that in $B(L^2(M, E \otimes \mathcal{A}))$

$$\|\mathcal{K} - S(D(\rho)^2 - \lambda)\| \leq \tfrac{1}{2}.$$

This choice is possible since by assumption $(D(\rho)^2 - \lambda)$ has a bounded inverse, hence also $(D(\rho)^2 - \bar{\lambda})$ has a bounded inverse on $L^2(M, E \otimes \mathcal{A})$.

It follows that

$$(Q(\lambda) + S)(D(\rho)^2 - \lambda) = 1 - \bigl(\mathcal{K} - S(D(\rho)^2 - \lambda)\bigr)$$

has a bounded inverse on $L^2(M, E \otimes \mathcal{A})$.

Prop. 5.3.1 implies that $1 - \mathcal{K} - S(D(\rho)^2 - \lambda)$ is invertible on $L^2(M, E \otimes \hat{\Omega}_{\leq \mu} \mathcal{A}_i)$ as well. Thus

$$\bigl(1 - (\mathcal{K} - S(D(\rho)^2 - \lambda))\bigr)^{-1}(Q(\lambda) + S)$$

is a bounded operator on $L^2(M, E \otimes \hat{\Omega}_{\leq\mu}\mathcal{A}_i)$, which is a right inverse for $(D(\rho)^2 - \lambda)$. Hence $D(\rho)^2 - \lambda$ is injective and bounded below. It remains to show that its range is dense. This follows from the fact that $D(\rho)^2 - \lambda$ is invertible on $L^2(M, E \otimes \mathcal{A})$.

Since $Q(\lambda)$ acts continuously on $\mathcal{S}_R(M, E \otimes \hat{\Omega}_{\leq\mu}\mathcal{A}_i)$ and $\mathcal{K}$ maps $L^2(M, E \otimes \hat{\Omega}_{\leq\mu}\mathcal{A}_i)$ continuously to $\mathcal{S}_R(M, E \otimes \hat{\Omega}_{\leq\mu}\mathcal{A}_i)$ by the previous lemma, the operator $(D(\rho)^2 - \lambda)^{-1}$ acts continuously on $\mathcal{S}_R(M, E \otimes \hat{\Omega}_{\leq\mu}\mathcal{A}_i)$ by

$$\begin{aligned}(D(\rho)^2 - \lambda)^{-1} &= (1 - \mathcal{K})(D(\rho)^2 - \lambda)^{-1} + \mathcal{K}(D(\rho)^2 - \lambda)^{-1} \\ &= Q(\lambda) + \mathcal{K}(D(\rho)^2 - \lambda)^{-1} \ .\end{aligned}$$

□

PROPOSITION 3.5.3. *Let $\rho \in \mathbb{R}$. Let $\lambda \in \mathbb{C}$ with $\operatorname{Re}\lambda < \omega$ be such that $(D(\rho)^2 - \lambda)$ has a bounded inverse on $L^2(M, E \otimes \hat{\Omega}_{\leq\mu}\mathcal{A}_i)$.*

Then for $\nu \in \mathbb{N}$, $\nu \geq 2$, the operator $(D(\rho)^2 - \lambda)^{-\nu}$ maps $L^2(M, E \otimes \hat{\Omega}_{\leq\mu}\mathcal{A}_i)$ continuously to $C_R^{2\nu-3}(M, E \otimes \hat{\Omega}_{\leq\mu}\mathcal{A}_i)$.

PROOF. Let $Q(\lambda)(D(\rho)^2 - \lambda)^{\nu} = 1 - \mathcal{K}$ as before, thus

$$(D(\rho)^2 - \lambda)^{-\nu} = Q(\lambda) + \mathcal{K}(D(\rho)^2 - \lambda)^{-\nu} \ .$$

By Prop. 3.4.3 and Lemma 5.2.18 the operator $Q(\lambda)$ maps $L^2(M, E \otimes \hat{\Omega}_{\leq\mu}\mathcal{A}_i)$ continuously to $C_R^{2\nu-3}(M, E \otimes \hat{\Omega}_{\leq\mu}\mathcal{A}_i)$. Furthermore $\mathcal{K}$ is smoothing. □

COROLLARY 3.5.4. *The kernel of $D(\rho)^2$ on $L^2(M, E \otimes \hat{\Omega}_{\leq\mu}\mathcal{A}_i)$ is a subspace of $\mathcal{S}_R(M, E \otimes \hat{\Omega}_{\leq\mu}\mathcal{A}_i)$.*

PROOF. Let $\lambda \neq 0$ be as in the previous proposition.

Then $(D(\rho)^2 - \lambda)^{-\nu}f = (-\lambda)^{-\nu}f$ for $f \in \operatorname{Ker} D(\rho)^2$ and every $\nu \in \mathbb{N}$. By the previous proposition it follows that the elements of $\operatorname{Ker} D(\rho)^2$ are smooth.

For $k \in \mathbb{Z}/6$ and $f \in \operatorname{Ker} D(\rho)^2$

$$D_{Z_k}^2 \phi_k f \in C_{cR}^\infty(Z_k, (\hat{\Omega}_{\leq\mu}\mathcal{A}_i)^{4d}) \ .$$

From Cor. 3.4.2 it follows that

$$\phi_k f = D_{Z_k}^{-2}(D_{Z_k}^2 \phi_k f) \in \mathcal{S}_R(Z_k, (\hat{\Omega}_{\leq\mu}\mathcal{A}_i)^{4d}) \ .$$

Hence $f \in \mathcal{S}_R(M, E \otimes \hat{\Omega}_{\leq\mu}\mathcal{A}_i)$. □

PROPOSITION 3.5.5. *Let $\rho \neq 0$.*

Let P be the projection onto the kernel of $D(\rho)$ on $L^2(M, E \otimes \mathcal{A})$.

Then P is a finite Hilbert-Schmidt operator whose integral kernel is of the form $\sum_{j=1}^m f_j(x)h_j(y)^*$ with $f_j, h_j \in \operatorname{Ker} D(\rho) \cap \mathcal{S}_R(M, E \otimes \mathcal{A}_\infty)$.

Furthermore on $L^2(M, E \otimes \hat{\Omega}_{\leq\mu}\mathcal{A}_i)$ we have that $\operatorname{Ker} D(\rho)^2 = PL^2(M, E \otimes \hat{\Omega}_{\leq\mu}\mathcal{A}_i)$ and $\operatorname{Ran} D(\rho)^2 = (1-P)L^2(M, E \otimes \hat{\Omega}_{\leq\mu}\mathcal{A}_i)$. Hence there is a decomposition

$$L^2(M, E \otimes \hat{\Omega}_{\leq\mu}\mathcal{A}_i) = \operatorname{Ker} D(\rho)^2 \oplus \operatorname{Ran} D(\rho)^2$$

with respect to which

$$D(\rho)^2 = 0 \oplus D(\rho)^2|_{\operatorname{Ran} D(\rho)^2} \ .$$

Moreover $D(\rho)^2|_{\operatorname{Ran} D(\rho)^2}$ is invertible.

PROOF. First consider the situation on $L^2(M, E \otimes \mathcal{A})$: Since the range of $D(\rho)$ is closed, there is an orthogonal projection P onto the kernel of $D(\rho)$ by Prop. 5.1.12. Furthermore $D(\rho)$ is selfadjoint, hence $\operatorname{Ker} D(\rho) = \operatorname{Ker} D(\rho)^2$. The range of $D(\rho)^2$ is closed, thus zero is an isolated point in the spectrum of $D(\rho)^2$ on $L^2(M, E \otimes \mathcal{A})$.

Hence, for r small enough,
$$P = \frac{1}{2\pi i} \int_{|\lambda|=r} (D(\rho)^2 - \lambda)^{-1} d\lambda .$$

From Prop. 3.5.2 it follows that zero is an isolated point in the spectrum of $D(\rho)^2$ on $L^2(M, E \otimes \hat{\Omega}_{\leq \mu} \mathcal{A}_i)$ as well. Thus P is well-defined as a bounded operator on $L^2(M, E \otimes \hat{\Omega}_{\leq \mu} \mathcal{A}_i)$. By Prop. 5.3.6 it is a Hilbert-Schmidt operator whose integral kernel is as asserted.

The remaining parts follow from the spectral theory for closed operators on Banach spaces ([**Da**], Th. 2.14). $\square$

COROLLARY 3.5.6. *Let* $\operatorname{Re} \lambda < \omega$.
(1) *Let* $\rho \neq 0$ *and let* P *be the orthogonal projection onto the kernel of* $D(\rho)^2$. *If* $D(\rho)^2 + P - \lambda$ *has a bounded inverse on* $L^2(M, E \otimes \mathcal{A})$, *then* $D(\rho)^2 + P - \lambda$ *has a bounded inverse on* $L^2(M, E \otimes \hat{\Omega}_{\leq \mu} \mathcal{A}_i)$ *and the inverse acts as a bounded operator on the space* $\mathcal{S}_R(M, E \otimes \hat{\Omega}_{\leq \mu} \mathcal{A}_i)$ *as well.*
(2) *Let* P_0 *be the orthogonal projection onto* $\operatorname{Ker} D_s^2$. *If* $D_s^2 + P_0 - \lambda$ *has a bounded inverse on* $L^2(M, E \otimes \mathcal{A})$, *then* $D_s^2 + P_0 - \lambda$ *has a bounded inverse on* $L^2(M, E \otimes \hat{\Omega}_{\leq \mu} \mathcal{A}_i)$ *and the inverse acts as a bounded operator on the space* $\mathcal{S}_R(M, E \otimes \hat{\Omega}_{\leq \mu} \mathcal{A}_i)$ *as well.*

In particular there is $c > 0$ *such that* $\{\operatorname{Re} \lambda < c\}$ *is in the resolvent set of* $D(\rho)^2 + P$ *resp.* $D_s^2 + P_0$.

PROOF. (1) From the previous proposition it follows that $P(1 - \lambda)^{-1} + (1 - P)(D(\rho)^2 - \lambda)^{-1}$ inverts $D(\rho)^2 + P - \lambda$ on $L^2(M, E \otimes \hat{\Omega}_{\leq \mu} \mathcal{A}_i)$. Since P acts as a bounded operator on the space $\mathcal{S}_R(M, E \otimes \hat{\Omega}_{\leq \mu} \mathcal{A}_i)$ by Cor. 3.5.4 and $(D(\rho)^2 - \lambda)^{-1}$ is bounded on $\mathcal{S}_R(M, E \otimes \hat{\Omega}_{\leq \mu} \mathcal{A}_i)$ by Prop. 3.5.2, the operator $(D(\rho)^2 + P - \lambda)^{-1}$ is bounded on $\mathcal{S}_R(M, E \otimes \hat{\Omega}_{\leq \mu} \mathcal{A}_i)$ as well.

(2) follows analogously. $\square$

3.5.3. An approximation of the semigroup. By cutting and pasting we construct a family of integral operators that behaves similar to a semigroup generated by $-D(\rho)^2$ for small times.

We work in the setting fixed in §3.5.1.

Let $e(\rho)_t^{\clubsuit}(x, y)$ be the restriction of the integral kernel of $e^{-t(D_N + \rho K)^2}$ to $\mathcal{U}_{\clubsuit} \times \mathcal{U}_{\clubsuit}$ and for $k \in \mathbb{Z}/6$ let $e(\rho)_t^k(x, y)$ be the restriction of the integral kernel of $e^{-tD_{Z_k}^2}$ to $\mathcal{U}_k \times \mathcal{U}_k$. Extend these functions by zero to $M \times M$. Clearly for $k \in \mathbb{Z}/6$ we have $e(\rho)_t^k(x, y) = e(0)_t^k(x, y)$.

We write $E(\rho)_t$ for the family of integral operators on $L^2(M, E \otimes \hat{\Omega}_{\leq \mu} \mathcal{A}_i)$ corresponding to the integral kernel
$$e(\rho)_t(x, y) := \sum_{k \in J} \gamma_k(x) e(\rho)_t^k(x, y) \phi_k(y)$$
and set $E(\rho)_0 := 1$.

Using the results in §3.2 and §3.4 one deduces:
(1) The family $E(\rho)_t$ is strongly continuous in t on $L^2(M, E \otimes \hat{\Omega}_{\leq\mu}\mathcal{A}_i)$.
(2) If $f \in C^\infty_{cR}(M, E\otimes\hat{\Omega}_{\leq\mu}\mathcal{A}_i)$, then the map $[0,\infty) \to L^2(M, E\otimes\hat{\Omega}_{\leq\mu}\mathcal{A}_i)$, $t \mapsto E(\rho)_t f$ is differentiable.
(3) $\operatorname{Ran} E(\rho)_t \subset \mathcal{S}_R(M, E \otimes \hat{\Omega}_{\leq\mu}\mathcal{A}_i)$ for $t > 0$.
(4) Let A be a differential operator on $C^\infty(M, E \otimes \hat{\Omega}_{\leq\mu}\mathcal{A}_i)$ of order m with bounded coefficients. For $T > 0$ there is $C > 0$ such that for all $0 < t < T$ on $L^2(M, E \otimes \hat{\Omega}_{\leq\mu}\mathcal{A}_i)$

$$\|AE(\rho)_t\| \leq Ct^{-\frac{m}{2}} .$$

(5) For any $m \in \mathbb{N}_0$ and $T > 0$ there is $C > 0$ such that for all $y \in \mathcal{U}_\clubsuit$, $\rho \in [-1,1]$ and $0 < t < T$ in $C^m(\mathcal{U}_\clubsuit, (E \otimes E_y) \otimes \mathcal{A}_i)$

$$\|e(\rho)^\clubsuit_t(\cdot, y) - e(0)^\clubsuit_t(\cdot, y)\|_{C^m} \leq Ct|\rho| .$$

An analogous estimate holds for the partial derivatives in y with respect to unit vector fields on $\mathcal{U}_\clubsuit$.

The last statement follows by Volterra development (Prop. 5.4.4): On N

$$e^{-tD_N(\rho)^2} - e^{-tD_N^2}$$
$$= \rho t \sum_{n=1}^\infty (-1)^n (\rho t)^{n-1} \int_{\Delta^n} e^{-u_0 t D_N^2} \big([D_N, K]_s + \rho K^2\big) e^{-u_1 t D_N^2} \ldots$$
$$\ldots e^{-u_n t D_N^2} \, du_0 \ldots du_n ,$$

and the sum is an integral operator whose integral kernel is uniformly bounded in $0 < t < T$ and $\rho \in [-1,1]$.

3.5.4. The semigroup $e^{-tD_s^2}$. By Cor. 3.1.5 the operator $e^{-tD_s^2}$ on the Hilbert space $L^2(M,E)$ is an integral operator with smooth integral kernel k_t for $t > 0$. In this section we show that k_t defines a strongly continuous semigroup on $L^2(M, E \otimes \hat{\Omega}_{\leq\mu}\mathcal{A}_i)$ and that this semigroup extends to a holomorphic one.

For D_s define $e^k_t(x,y)$ analogously to $e(\rho)^k_t(x,y)$ for $D(\rho)$ in the previous section and let

$$e_t(x,y) := \sum_{k \in J} \gamma_k(x) e^k_t(x,y) \phi_k(y) .$$

The corresponding family of operators is denoted by E_t. We set $E_0 := 1$. The properties of E_t are as described in the previous section.

For $f \in C^\infty_{cR}(M,E)$

$$e^{-tD_s^2} f - E_t f = -\int_0^t e^{-sD_s^2} \left(\frac{d}{dt} + D_s^2\right) E_{t-s} f \, ds$$

by Duhamel's principle.

PROPOSITION 3.5.7. (1) *The heat kernel k_t associated to D_s defines a strongly continuous semigroup on $L^2(M, E\otimes\hat{\Omega}_{\leq\mu}\mathcal{A}_i)$ with generator $-D_s^2$, which extends to a bounded holomorphic semigroup.*
(2) *Let A be a differential operator of order $m \in \mathbb{N}_0$ with bounded smooth coefficients. Then for any $t > 0$ the operator $Ae^{-tD_s^2}$ is bounded on*

3.5. THE HEAT SEMIGROUP ON M

$L^2(M, E \otimes \hat{\Omega}_{\leq \mu} \mathcal{A}_i)$ and for any $T > 0$ there is $C > 0$ such that for $0 < t < T$

$$\|Ae^{-tD_s^2}\| \leq Ct^{-\frac{m}{2}} .$$

PROOF. First we show that for $T > 0$ there is $C > 0$ such that for $0 < t < T$ the difference $A_x k_t(x,y) - A_x e_t(x,y)$ is bounded by Ct in $L^2(M \times M, E \boxtimes E^*)$.

For $k \in J$ let $\chi_k \in C_c^\infty(M)$ be a function with values in $[0,1]$, with compact support in $\mathcal{U}_k$ and equal to one on a neighborhood of $\operatorname{supp} d\gamma_k$.

From Duhamel's principle it follows that

$$A_x k_t(x,y) - A_x e_t(x,y)$$
$$= -\sum_{k \in J} \int_0^t \int_M A_x k_s(x,r)[(\partial\!\!\!/_E)_r, c(d\gamma_k(r))]_s e_{t-s}^k(r,y)\phi_k(y) \, drds .$$

This can be re-written as

$$-\sum_{k \in J} \int_0^t \int_M (1-\chi_k(x))A_x k_s(x,r)[(\partial\!\!\!/_E)_r, c(d\gamma_k(r))]_s e_{t-s}^k(r,y)\phi_k(y) \, drds$$

$$-\sum_{k \in J} \int_0^t \int_M \chi_k(x)A_x(k_s(x,r) - e_s^k(x,r))[(\partial\!\!\!/_E)_r, c(d\gamma_k(r))]_s e_{t-s}^k(r,y)\phi_k(y) \, drds$$

$$-\sum_{k \in J} \int_0^t \int_M \chi_k(x)A_x e_s^k(x,r)[(\partial\!\!\!/_E)_r, c(d\gamma_k(r))]_s e_{t-s}^k(r,y)\phi_k(y) \, drds .$$

Using Lemma 3.1.8 and Lemma 3.1.9 we estimate the norms of the three terms in $E_x \otimes E_y$:

Since $\operatorname{supp} d\gamma_k \cap \operatorname{supp} \phi_k = \emptyset$ and $\operatorname{supp} d\gamma_k \cap \operatorname{supp}(1-\chi_k) = \emptyset$, there is $C > 0$ such that for $x, y \in M$ and $t > 0$ the norm of the first term is bounded by

$$C \sum_{k \in J} t(1-\chi_k(x)) e^{-\frac{d(x,\operatorname{supp} d\gamma_k)^2}{(4+\delta)t}} e^{-\frac{d(y,\operatorname{supp} d\gamma_k)^2}{(4+\delta)t}} 1_{\operatorname{supp} \phi_k}(y) ,$$

and such that for $x, y \in M$ and $t > 0$ the norms of the second and third term are bounded by

$$Ct\chi_k(x) e^{-\frac{d(y,\operatorname{supp} d\gamma_k)^2}{(4+\delta)t}} 1_{\operatorname{supp} \phi_k}(y) .$$

When estimating the third term we used that the action of the integral kernel $e_s^k(x,r)\chi_k(r)$ is uniformly bounded for $k = \clubsuit$ on $C^n(\mathcal{U}_\clubsuit, E \otimes E_y)$ by Prop. 3.2.2 and for $k \in \mathbb{Z}/6$ on $C_R^n(\mathcal{U}_k, E \otimes E_y)$ by Prop. 3.4.1.

Analogous estimates hold for the derivatives in y with respect to unit vector fields on M.

Hence $A_x k_t(x,y) - A_x e_t(x,y)$ is bounded by Ct in $L^2(M \times M, E \boxtimes E^*)$ for $0 < t < T$ and some $C > 0$. By Cor. 5.2.4 the corresponding family of operators on $L^2(M, E \otimes \hat{\Omega}_{\leq \mu} \mathcal{A}_i)$ is bounded by Ct for $0 < t < T$, hence $A_x k_t(x,y)$ defines a family of bounded operators on $L^2(M, E \otimes \hat{\Omega}_{\leq \mu} \mathcal{A}_i)$.

Write $S(t)$ for the integral operator induced by the integral kernel $k_t(x,y)$.

By property (4) in §3.5.3 there is $C > 0$ such that $\|AE_t\| \leq Ct^{-\frac{m}{2}}$ on $L^2(M, E \otimes \hat{\Omega}_{\leq \mu} \mathcal{A}_i)$ for $0 < t < T$ and some $C > 0$, hence

$$(*) \quad \|AS(t)\| \leq Ct^{-\frac{m}{2}} .$$

The fact that $S(t)$ extends to a bounded holomorphic semigroup on $L^2(M, E \otimes \hat{\Omega}_{\leq \mu} \mathcal{A}_i)$ is an almost immediate consequence of $(*)$:

Since E_t converges strongly to the identity on $L^2(M, E \otimes \hat\Omega_{\le\mu}\mathcal{A}_i)$ for $t \to 0$, the operator $S(t)$ also does. Furthermore the kernels k_t obey the semigroup law, hence $S(t)$ is a strongly continuous semigroup on $L^2(M, E \otimes \hat\Omega_{\le\mu}\mathcal{A}_i)$.

Note that the range of $S(t) - E_t$ is a subset of $\mathcal{S}_R(M, E \otimes \hat\Omega_{\le\mu}\mathcal{A}_i)$. Hence $\mathcal{S}_R(M, E \otimes \hat\Omega_{\le\mu}\mathcal{A}_i)$ is invariant under the action of $S(t)$. It follows that $-D_s^2$ is the generator of $S(t)$.

From Prop. 5.4.3 and the estimate $(*)$ applied to $A = D_s^2$ it follows that the semigroup $S(t) = e^{-tD_s^2}$ extends to a holomorphic semigroup on $L^2(M, E \otimes \hat\Omega_{\le\mu}\mathcal{A}_i)$.

In order to show that the holomorphic semigroup is bounded, let P_0 be the projection onto the kernel of D_s^2.

By Cor. 3.5.6 there is $c > 0$ such that $\{\operatorname{Re}\lambda < c\}$ is in the resolvent set of $D_s^2 + P_0$ on $L^2(M, E \otimes \hat\Omega_{\le\mu}\mathcal{A}_i)$. Hence by Prop. 5.4.2 the holomorphic semigroup $e^{-t(D_s^2 + P_0)}$ is bounded. Thus

$$e^{-tD_s^2} = e^{-t(D_s^2 + P_0)}(1 - P_0) + P_0$$

is bounded as well. $\square$

Recall that in §3.5.1 we fixed the domain of D_s^2 as an unbounded operator on $L^2(M, E \otimes \hat\Omega_{\le\mu}\mathcal{A}_i)$, but not the domain of D_s. This is done now:

Let D_s be the closure on $L^2(M, E \otimes \hat\Omega_{\le\mu}\mathcal{A}_i)$ of the Dirac operator $\partial\!\!\!/_E$ with domain $\mathcal{S}_R(M, E \otimes \hat\Omega_{\le\mu}\mathcal{A}_i)$.

COROLLARY 3.5.8. *Let P_0 be the projection onto the kernel of D_s^2. Let $\lambda \in \mathbb{C}$ with $\operatorname{Re}\lambda^2 < 0$.*

Then the operators $D_s + P_0$ and $D_s - \lambda$ have a bounded inverse on $L^2(M, E \otimes \hat\Omega_{\le\mu}\mathcal{A}_i)$.

PROOF. By Cor. 3.5.6 there is $c > 0$ such that $\{\operatorname{Re}\lambda \le c\}$ is in the resolvent set of $D_s^2 + P_0$ on $L^2(M, E \otimes \hat\Omega_{\le\mu}\mathcal{A}_i)$.

By the previous proposition and Prop. 5.4.2 it follows that there is $C > 0$ such that for all $t > 0$ on $L^2(M, E \otimes \hat\Omega_{\le\mu}\mathcal{A}_i)$

$$\|(D_s + P_0)e^{-t(D_s^2 + P_0)}\| \le C t^{-\frac{1}{2}} e^{-ct} .$$

Thus

$$G := \int_0^\infty (D_s + P_0) e^{-t(D_s^2 + P_0)} dt$$

is a bounded operator on $L^2(M, E \otimes \hat\Omega_{\le\mu}\mathcal{A}_i)$. Furthermore $G = (D_s + P_0)^{-1}$ on the Hilbert space $L^2(M, E)$. From Cor. 3.5.6 it follows that G acts as a bounded operator on $\mathcal{S}_R(M, E)$. Hence G inverts $D_s + P_0$ on $\mathcal{S}_R(M, E) \odot \hat\Omega_{\le\mu}\mathcal{A}_i$, thus G inverts $D_s + P_0$ on $L^2(M, E \otimes \hat\Omega_{\le\mu}\mathcal{A}_i)$.

The proof of the fact that

$$\int_0^\infty (D_s + \lambda) e^{-t(D_s^2 - \lambda^2)}$$

inverts $(D_s - \lambda)$ for $\operatorname{Re}\lambda^2 < 0$ is analogous. $\square$

This answers a question from the beginning of §3.5.1: the operator D_s^2 is indeed the square of D_s.

3.5.5. The semigroup $e^{-tD(\rho)^2}$. This section is devoted to the study of the holomorphic semigroup generated by $-D(\rho)^2$.

First we prove its existence.

Define $D(\rho)$ on $L^2(M, E \otimes \hat{\Omega}_{\leq \mu}\mathcal{A}_i)$ as the closure of the operator $\bar{\partial}_E + \rho K$ with domain $\mathcal{S}_R(M, E \otimes \hat{\Omega}_{\leq \mu}\mathcal{A}_i)$.

Let P_0 be the orthogonal projection onto the kernel of D_s^2. Then $D(\rho)$ is a bounded perturbation of $W^*(D_s + P_0)W$ with W as in §2.1.2. By the results of the previous section we can apply Prop. 5.4.10 and conclude that $D(\rho)^2$, as defined in §3.5.1, is the square of $D(\rho)$ and furthermore that $-D(\rho)^2$ generates a holomorphic semigroup. This shows the first assertion of the following proposition.

For $\rho \neq 0$ let P be the orthogonal projection onto the kernel of $D(\rho)^2$.

PROPOSITION 3.5.9.   (1) *Let $\rho \in \mathbb{R}$. The operator $-D(\rho)^2$ generates a holomorphic semigroup $e^{-tD(\rho)^2}$ on $L^2(M, E \otimes \hat{\Omega}_{\leq \mu}\mathcal{A}_i)$. For $\rho \neq 0$ the semigroup is bounded holomorphic.*
(2) *For $t \geq 0$ the operator $e^{-tD(\rho)^2}$ depends analytically on ρ. For every $T > 0$ there is $C > 0$ such that*
$$\|e^{-tD(\rho)^2}\| \leq C$$
for all $\rho \in [-1, 1]$ and $0 \leq t \leq T$.
(3) *For $\rho \neq 0$ there are $C, \omega > 0$ such that for all $t \geq 0$*
$$\|(1 - P)e^{-tD(\rho)^2}\| \leq C e^{-\omega t} \ .$$

PROOF. (1) Let $\rho \neq 0$. By Cor. 3.5.6 there is $c > 0$ such that $\{\text{Re}\,\lambda \leq c\}$ is in the resolvent set of $D(\rho)^2 + P$ on $L^2(M, E \otimes \hat{\Omega}_{\leq \mu}\mathcal{A}_i)$, hence by Lemma 5.4.2 the holomorphic semigroup $e^{-t(D(\rho)^2 + P)}$ is bounded by Ce^{-ct} for some $C > 0$, thus for $T > 0$ there is $C > 0$ such that for $t > T$
$$\|D(\rho)^2 e^{-tD(\rho)^2}\| = \|D(\rho)^2 e^{-t(D(\rho)^2 + P)}\| \leq C e^{-ct} \ .$$
Now Prop. 5.4.3 implies that the semigroup $e^{-tD(\rho)^2}$ is bounded holomorphic.

(2) follows from Prop. 5.4.4.

(3) follows from $(1 - P)e^{-tD(\rho)^2} = (1 - P)e^{-t(D(\rho)^2 + P)}$. □

The following proposition shows that $e^{-tD(\rho)^2}$ is smoothing.

PROPOSITION 3.5.10.   (1) *Let $n \in \mathbb{N}_0$. For every $\rho \in \mathbb{R}$ and every $t > 0$ the operator $e^{-tD(\rho)^2}$ maps $L^2(M, E \otimes \hat{\Omega}_{\leq \mu}\mathcal{A}_i)$ continuously to $C_R^n(M, E \otimes \hat{\Omega}_{\leq \mu}\mathcal{A}_i)$.*
(2) *Let $n \in 2\mathbb{N}$, $n \geq 4$. For every $\rho \neq 0$ the family $e^{-tD(\rho)^2} : C_{cR}^n(M, E \otimes \hat{\Omega}_{\leq \mu}\mathcal{A}_i) \to C_R^{n-3}(M, E \otimes \hat{\Omega}_{\leq \mu}\mathcal{A}_i)$ is uniformly bounded.*
(3) *Let $n \in 2\mathbb{N}$, $n \geq 4$. For every $T > 0$ the family $e^{-tD(\rho)^2} : C_{cR}^n(M, E \otimes \hat{\Omega}_{\leq \mu}\mathcal{A}_i) \to C_R^{n-3}(M, E \otimes \hat{\Omega}_{\leq \mu}\mathcal{A}_i)$ is uniformly bounded in $0 \leq t < T$ and in $\rho \in [-1, 1]$.*

PROOF. We conclude (1) from
$$e^{-tD(\rho)^2} = (D(\rho)^2 + 1)^{-k}(D(\rho)^2 + 1)^k e^{-tD(\rho)^2} \ ,$$
taking into account that $(D(\rho)^2 + 1)^k e^{-tD(\rho)^2}$ is bounded on $L^2(M, E \otimes \hat{\Omega}_{\leq \mu}\mathcal{A}_i)$ for $t > 0$ and that $(D(\rho)^2 + 1)^{-k}$ maps $L^2(M, E \otimes \hat{\Omega}_{\leq \mu}\mathcal{A}_i)$ continuously to $C_R^{2k-3}(M, E \otimes \hat{\Omega}_{\leq \mu}\mathcal{A}_i)$ for $k \in \mathbb{N}$, $k \geq 2$, by Prop. 3.5.3.

(2) and (3) follow from
$$e^{-tD(\rho)^2} = (D(\rho)^2 + 1)^{-k} e^{-tD(\rho)^2} (D(\rho)^2 + 1)^k$$
by Prop. 3.5.3. $\square$

3.5.6. The integral kernel. In order to prove that the operator $e^{-tD(\rho)^2}$ is an integral operator we use the same method as in §3.3.2: Via Duhamel's principle we compare $e^{-tD(\rho)^2}$ and the approximation $E(\rho)_t$, which was defined in §3.5.3.

In the following $|\cdot|$ denotes the norm on the fibers of $(E \boxtimes E^*) \otimes \mathcal{A}_i$.

PROPOSITION 3.5.11. *For every $\rho \in \mathbb{R}$ and every $t > 0$ the operator $e^{-tD(\rho)^2}$ is an integral operator with smooth integral kernel. Let $k(\rho)_t(x, y)$ be its integral kernel.*

(1) *The map $(0, \infty) \to C^\infty(M \times M, (E \boxtimes E^*) \otimes \mathcal{A}_\infty)$, $t \mapsto k(\rho)_t$ is smooth.*
(2) $k(\rho)_t(x, y) = k(\rho)_t(y, x)^*$.
(3) *For every $T > 0$ there are $c, C > 0$ such that*
$$|k(\rho)_t(x,y) - e(\rho)_t(x,y)| \leq Ct\big(|\rho| 1_{\mathcal{U}_\clubsuit}(y) + \sum_{k \in J} e^{-\frac{d(y, \text{supp } d\gamma_k)^2}{ct}} 1_{\text{supp } \phi_k}(y)\big)$$
for all $0 < t < T$, $\rho \in [-1, 1]$ and $x, y \in M$.
(4) *Let $\rho \neq 0$. There are $c, C > 0$ such that*
$$|k(\rho)_t(x,y) - e(\rho)_t(x,y)| \leq Cte^{-\frac{d(y, \mathcal{U}_\clubsuit)^2}{ct}}$$
for all $t > 0$ and all $x, y \in M$.

Statements analogous to (3) and (4) hold for the partial derivatives in x and y with respect to unit vector fields on M.

PROOF. For (1) it is enough to prove an analogous statement for the operator $e^{-tD(\rho)^2} - E(\rho)_t$.

Let $f \in C^\infty_{cR}(M, E \otimes \mathcal{A}_i)$. Then by Duhamel's principle
$$e^{-tD(\rho)^2} f - E(\rho)_t f$$
$$= -\sum_{k \in J} \int_0^t \int_M e^{-sD(\rho)^2} \big(\frac{d}{dt} + D(\rho)^2\big) \gamma_k e(\rho)^k_{t-s}(\cdot, y) \phi_k(y) f(y) \, dy \, ds \; .$$

We write
$$\sum_{k \in J} \big(\frac{d}{d\tau} + D(\rho)^2\big) \gamma_k e(\rho)^k_\tau(\cdot, y) \phi_k(y)$$
$$= -\sum_{k \in J} [\partial\!\!\!/_E, c(d\gamma_k)]_s e(0)^k_\tau(\cdot, y) \phi_k(y)$$
$$- [\partial\!\!\!/_E, c(d\gamma_\clubsuit)]_s \big(e(\rho)^\clubsuit_\tau(\cdot, y) - e(0)^\clubsuit_\tau(\cdot, y)\big) \phi_\clubsuit(y) \; .$$

Note that the map
$$(0, \infty) \to C^\infty(M, C^\infty_{cR}(M, E \otimes \mathcal{A}_\infty) \otimes E) \; ,$$
$$\tau \mapsto \big(y \mapsto \sum_{k \in J} \big(\frac{d}{d\tau} + D(\rho)^2\big) e(\rho)^k_\tau(\cdot, y) \phi_k(y)\big)$$
is smooth. We show that it extends smoothly by zero to $\tau = 0$. From this and Prop. 3.5.10 it follows that $e^{-tD(\rho)^2} - E(\rho)_t$ is an integral operator.

We estimate the terms on the right hand side of the previous equation.

From the estimates in Lemma 3.1.8 and Lemma 3.4.4 it follows that for any $m \in \mathbb{N}_0$ there are $C, c > 0$ such that

$$\|[\bar{\partial}_E, c(d\gamma_k)]_s e(0)^k_\tau(\cdot, y)\phi_k(y)\|_{C^m_R} \leq C e^{-\frac{d(y, \text{supp } d\gamma_k)^2}{c\tau}} 1_{\text{supp } \phi_k}(y)$$

in $C^m_R(M, E \otimes \mathcal{A}_i) \otimes E^*_y$ for all $k \in J$, $0 < \tau < T$, $\rho \in [-1, 1]$ and $y \in M$.

Furthermore by §3.5.3, property (5), there is $C > 0$ such that

$$\|[\bar{\partial}_E, c(d\gamma_\clubsuit)]_s \bigl(e(\rho)^\clubsuit_\tau(\cdot, y) - e(0)^\clubsuit_\tau(\cdot, y)\bigr)\phi_\clubsuit(y)\|_{C^m_R} \leq C\tau|\rho|1_{\mathcal{U}_\clubsuit}(y)$$

in $C^m_R(M, E \otimes \mathcal{A}_i) \otimes E_y$ for $0 < \tau < T$, $\rho \in [-1, 1]$ and $y \in M$.

Analogous estimates hold for the derivatives in y and also in τ since by the heat equation the derivatives with respect to τ can be expressed in terms of the derivatives with respect to x. This shows (1).

The property $k(\rho)_t(x, y) = k(\rho)_t(y, x)^*$ follows from the selfadjointness of $e^{-tD(\rho)^2}$.

Statement (2) and (3) follow from the estimates using Prop. 3.5.10. For the proof of (3) we also take into account that the kernel $e(\rho)^\clubsuit_t$ and its derivatives are uniformly bounded in t with $t > T$. □

COROLLARY 3.5.12. *Let $\nu \in \mathbb{N}_0$. For every $\rho \in \mathbb{R}$ and $T > 0$ there is $C > 0$ such that*

$$\|D(\rho)^\nu e^{-tD(\rho)^2}\| \leq Ct^{-\frac{\nu}{2}}$$

for $0 < t < T$ on $L^2(M, E \otimes \hat{\Omega}_{\leq \mu} \mathcal{A}_i)$.

PROOF. By Duhamel's principle, for $f \in C^\infty_{Rc}(M, E \otimes \hat{\Omega}_{\leq \mu} \mathcal{A}_i)$,

$$D(\rho)^\nu e^{-tD(\rho)^2} f - D(\rho)^\nu E(\rho)_t f$$
$$= -\sum_{k \in J} \int_0^t \int_M e^{-sD(\rho)^2} D(\rho)^\nu [D(\rho)^2, \gamma_k]_s e(\rho)^k_{t-s}(\cdot, y)\phi_k(y)f(y) \, dy \, ds \, .$$

There is $C > 0$ such that this term is bounded in $L^2(M, E \otimes \hat{\Omega}_{\leq \mu} \mathcal{A}_i)$ by Ct for $0 < t < T$. Furthermore by Prop. 3.2.2 and Prop. 3.4.1 there is $C > 0$ such that on $L^2(M, E \otimes \hat{\Omega}_{\leq \mu} \mathcal{A}_i)$

$$\|D(\rho)^\nu E(\rho)_t\| \leq Ct^{-\frac{\nu}{2}} \, .$$

The assertion follows. □

COROLLARY 3.5.13. *For every $\rho \neq 0$ and $m \in \mathbb{N}$ the family of integral kernels $k(\rho)_t$ defines a strongly continuous semigroup on $C^m_R(M, E \otimes \hat{\Omega}_{\leq \mu} \mathcal{A}_i)$ bounded by $C(1 + t)^{\frac{3}{2}}$ for some $C > 0$ and all $t > 0$.*

It is denoted by $e^{-tD(\rho)^2}$ as well.

PROOF. By the estimates in the proposition the integral kernel $k(\rho)_t - e(\rho)_t$ defines an operator on $C^m_R(M, E \otimes \hat{\Omega}_{\leq \mu} \mathcal{A}_i)$ bounded by $Ct^{3/2}$ for every $t > 0$ and some $C > 0$.

We show that $E(\rho)_t$ is a strongly continuous uniformly bounded family of operators on $C^m_R(M, E \otimes \hat{\Omega}_{\leq \mu} \mathcal{A}_i)$.

For $k \in \mathbb{Z}/6$ the action of the integral kernel $e(\rho)^k_t$ on $C^m_R(\mathcal{U}_k, E \otimes \hat{\Omega}_{\leq \mu} \mathcal{A}_i)$ is strongly continuous and uniformly bounded in t by Prop. 3.4.1.

By Prop. 3.2.3 and Prop. 5.4.4 the family $e^{-t(D_N+\rho K)^2}$ is a strongly continuous semigroup on $C^m(N, E_N \otimes \hat{\Omega}_{\leq \mu}\mathcal{A}_i)$. It is bounded since its integral kernel is uniformly bounded for $t > 1$. Hence the action of $e(\rho)_t^{\clubsuit}$ on $C^m(\mathcal{U}_{\clubsuit}, E \otimes \hat{\Omega}_{\leq \mu}\mathcal{A}_i)$ is strongly continuous and uniformly bounded.

Since $E(\rho)_t$ converges strongly to the identity on $C_R^m(\mathcal{U}_k, E \otimes \hat{\Omega}_{\leq \mu}\mathcal{A}_i)$ for $t \to 0$, so does $e^{-tD(\rho)^2}$. It clearly satisfies the semigroup property. $\square$

COROLLARY 3.5.14. *Let $\rho \neq 0$ and $n \in \mathbb{N}$.*

For every $\varepsilon > 0$ we can find $C, c > 0$ such that for $x, y \in M$ with $d(x,y) > \varepsilon$ and $t > 0$
$$|k(\rho)_t(x,y)| \leq C\left(e^{-\frac{d(x,y)^2}{ct}} + te^{-\frac{d(y,\mathcal{U}_{\clubsuit})^2}{ct}}\right).$$
An analogous statement holds for the derivatives in x and y with respect to unit vector fields on M.

We refer to §5.2.4 for the notion of a Hilbert-Schmidt operator and the Hilbert-Schmidt norm $\|\cdot\|_{HS}$ used in the next corollary.

COROLLARY 3.5.15. *Let $\rho \neq 0$ and $\nu \in \mathbb{N}_0$.*

The operators $1_{M_r}D(\rho)^\nu e^{-tD(\rho)^2}$ and $D(\rho)^\nu e^{-tD(\rho)^2} 1_{M_r}$ with $r \geq 0$ are Hilbert-Schmidt operators.

(1) *For every $T > 0$ there is $C > 0$ such that for any $r \geq 0$ and $t > T$*
$$\|1_{M_r}D(\rho)^\nu e^{-tD(\rho)^2}\|_{HS} \leq C(1+r)$$
and
$$\|D(\rho)^\nu e^{-tD(\rho)^2} 1_{M_r}\|_{HS} \leq C(1+r).$$

(2) *For every $\varepsilon > 0$ there is $C > 0$ such that for every $r, t > 0$*
$$\|1_{M_r}D(\rho)^\nu e^{-tD(\rho)^2}(1 - 1_{M_{r+\varepsilon}})\|_{HS} \leq C(1+r)t^{1/2}$$
and
$$\|(1 - 1_{M_{r+\varepsilon}})D(\rho)^\nu e^{-tD(\rho)^2} 1_{M_r}\|_{HS} \leq C(1+r)t^{1/2}.$$

PROOF. (1) Since by Prop. 3.5.9 the semigroup $e^{-tD(\rho)^2}$ is bounded, it follows from Prop. 5.2.13 that there is $C > 0$ such that for all $r > 0$ and $t > T$
$$\|1_{M_r}D(\rho)^\nu e^{-tD(\rho)^2}\|_{HS} \leq C\|1_{M_r}D(\rho)^\nu e^{-TD(\rho)^2}\|_{HS}.$$
By the previous corollary there are $C, c > 0$ such that
$$|1_{M_r}(x)D(\rho)^\nu_x k(\rho)_T(x,y)| \leq C 1_{M_r}(x)(e^{-cd(x,y)^2} + e^{-cd(y,M_r)^2})$$
for all $r > 0$ and $x, y \in M$. This yields the asserted estimate. The second estimate in (1) is proved analogously.

(2) follows from the previous corollary and (1). $\square$

CHAPTER 4

Superconnections and the Index Theorem

The notion of a superconnection on a free finitely generated $\mathbb{Z}/2$-graded module, which we define now, generalizes the notion of a connection on a free module [**Ka**].

In the family case superconnections usually act on infinite dimensional bundles ([**BGV**], Ch. 9). In analogy the superconnections we consider act on modules with infinitely many generators. In that sense the following definition should be merely seen as a motivation for the definitions of the superconnections to come.

DEFINITION 4.0.16. *Let $\mathcal{B}$ be a locally m-convex Fréchet algebra and let $p,q \in \mathbb{N}_0$.*

Let $V := (\mathbb{C}^+)^p \oplus (\mathbb{C}^-)^q$. Consider $V \otimes \hat{\Omega}_*\mathcal{B}$ as a $\mathbb{Z}/2$-graded space.

A SUPERCONNECTION *on $V \otimes \mathcal{B}$ is an odd linear map*

$$A : V \otimes \hat{\Omega}_*\mathcal{B} \to V \otimes \hat{\Omega}_*\mathcal{B}$$

satisfying Leibniz rule:
For $\alpha \in V^\pm \otimes \hat{\Omega}_k\mathcal{B}$ and $\beta \in \hat{\Omega}_*\mathcal{B}$

$$A(\alpha\beta) = A(\alpha)\beta + (-1)^{\deg \alpha}\alpha \, d\beta$$

where $\deg \alpha$ is the degree of α with respect to the $\mathbb{Z}/2$-grading of $V \otimes \hat{\Omega}_*\mathcal{B}$.

The map A^2 is called the CURVATURE *of A.*

As for a connection [**Ka**] the curvature of a superconnection is a right $\hat{\Omega}_*\mathcal{B}$-module map.

4.1. The superconnection A_t^I associated to D_I

4.1.1. The family $e^{-(A_t^I)^2}$. Let C_1 be the $\mathbb{Z}/2$-graded unital algebra generated by an odd element σ with $\sigma^2 = 1$. As a vector space C_1 is isomorphic to $\mathbb{C} \oplus \mathbb{C}$ via the map $C_1 \to \mathbb{C} \oplus \mathbb{C}, a + b\sigma \mapsto (a,b)$. We endow C_1 with the scalar product induced by the standard hermitian scalar product on $\mathbb{C} \oplus \mathbb{C}$.

We identify $L^2([0,1], C_1 \otimes (\hat{\Omega}_{\leq \mu}\mathcal{A}_i)^{2d})$ with $C_1 \otimes L^2([0,1], (\hat{\Omega}_{\leq \mu}\mathcal{A}_i)^{2d})$ and consider D_I, as defined in §3.3.1 with boundary conditions induced by a pair (P_0, P_1), as an unbounded operator on the $\mathbb{Z}/2$-graded $\mathcal{A}_i$-module $L^2([0,1], C_1 \otimes \mathcal{A}_i^{2d})$.

Let $U \in C^\infty([0,1], M_{2d}(\mathcal{A}_\infty))$ be as in Prop. 2.2.1 with $U(0)P_0U(0)^* = P_s$ and $U(1)P_1U(1)^* = 1 - P_s$. The map $U^*\,dU$ can be seen as a flat superconnection on $L^2([0,1], C_1 \otimes \mathcal{A}_i^{2d})$. It preserves the space $C_1 \otimes C_R^\infty([0,1], (\hat{\Omega}_{\leq \mu}\mathcal{A}_i)^{2d})$.

We define

$$A_I := U^*\,dU + \sigma D_I$$

and for $t \geq 0$

$$A_t^I := U^*\,dU + \sqrt{t}\sigma D_I \,.$$

Then A_I is an odd map on $C_1 \otimes C_R^\infty([0,1], (\hat{\Omega}_{\leq \mu} \mathcal{A}_i)^{2d})$ fulfilling Leibniz rule and is called a superconnection associated to σD_I. The map A_t^I is called the corresponding rescaled superconnection.

The curvature of A_I is

$$A_I^2 = U^* \, \mathrm{d}^2 U + U^* \, \mathrm{d} U \sigma D_I + \sigma D_I U^* \, \mathrm{d} U + D_I^2 = D_I^2 + \sigma[D_I, U^* \, \mathrm{d} U] \; .$$

From

$$\begin{aligned}[D_I, U^* \, \mathrm{d} U] &= -U^*[\mathrm{d}, U D_I U^*] U \\ &= -U^*([\mathrm{d}, D_{I_s}] + [\mathrm{d}, U I_0(\partial U^*)]) U \\ &= -U^* \, \mathrm{d}(U I_0(\partial U^*)) U =: R\end{aligned}$$

it follows that $A_I^2 = D_I^2 + \sigma R$ with $R \in C^\infty([0,1], M_{2d}(\hat{\Omega}_1 \mathcal{A}_\infty))$ vanishing near the boundary and fulfilling $R^* = -R$.

The curvature of the rescaled superconnection A_t^I is

$$(A_t^I)^2 = t D_I^2 + \sqrt{t} \sigma R \; .$$

We see that the curvature and the rescaled curvature are right $\hat{\Omega}_{\leq \mu} \mathcal{A}_i$-module maps.

Since A_I^2 is a bounded perturbation of D_I^2, it defines a holomorphic semigroup $e^{-t A_I^2}$ on $L^2([0,1], C_1 \otimes (\hat{\Omega}_{\leq \mu} \mathcal{A}_i)^{2d})$.

In the following we restrict to $t \geq 0$:

By Volterra development

$$\begin{aligned}e^{-t A_I^2} &= \sum_{n=0}^\infty (-1)^n t^n \int_{\Delta^n} e^{-u_0 t D_I^2} \sigma R e^{-u_1 t D_I^2} \sigma R \ldots e^{-u_n t D_I^2} \, du_0 \ldots du_n \\ &= \sum_{n=0}^\infty \sigma^n (-1)^{\frac{(n+1)n}{2}} t^n I_n(t)\end{aligned}$$

with

$$I_n(t) := \int_{\Delta^n} e^{-u_0 t D_I^2} R e^{-u_1 t D_I^2} R \ldots e^{-u_n t D_I^2} \, du_0 \ldots du_n \; .$$

Note that the series is finite on $L^2([0,1], C_1 \otimes (\hat{\Omega}_{\leq \mu} \mathcal{A}_i)^{2d})$.

It follows that

$$e^{-(A_t^I)^2} = \sum_{n=0}^\infty \sigma^n (-1)^{\frac{(n+1)n}{2}} t^{n/2} I_n(t) \; .$$

The operators $I_n(t)$ obey the following recursion relation for $n \geq 1$:

$$\begin{aligned}I_n(t) &= \int_0^1 du_0 \, e^{-u_0 t D_I^2} R \int_{(1-u_0)\Delta^{n-1}} e^{-u_1 t D_I^2} R \ldots R e^{-u_n t D_I^2} \, du_1 \ldots du_n \\ &= \int_0^1 du_0 \, (1-u_0)^{n-1} e^{-u_0 t D_I^2} R \int_{\Delta^{n-1}} e^{-(1-u_0) u_1 t D_I^2} R \ldots \\ &\qquad \ldots R e^{-(1-u_0) u_n t D_I^2} \, du_1 \ldots du_n \\ &= \int_0^1 du_0 \, (1-u_0)^{n-1} e^{-u_0 t D_I^2} R I_{n-1}((1-u_0)t) \; .\end{aligned}$$

Note that $e^{-(A_t^I)^2}$ is selfadjoint on $L^2([0,1], C_1 \otimes (\hat{\Omega}_{\leq \mu} \mathcal{A}_i)^{2d})$ in the sense of §5.2.3 and that $I_n(t) = (-1)^n I_n(t)^*$.

4.1.2. The integral kernel of $e^{-(A_t^I)^2}$**.** Since $e^{-tD_I^2}$ is a bounded semigroup on $C_R^m([0,1],(\hat{\Omega}_{\leq\mu}\mathcal{A}_i)^{2d})$ for every $m \in \mathbb{N}_0$ by Prop. 3.3.3, the family $I_n(t) : C_R^m([0,1],\mathcal{A}_i^{2d}) \to C_R^m([0,1],(\hat{\Omega}_n\mathcal{A}_i)^{2d})$ is uniformly bounded in $t \geq 0$.

In the following we write $|\cdot|$ for the norm on $M_{2d}(\hat{\Omega}_{\leq\mu}\mathcal{A}_i)$.

PROPOSITION 4.1.1. *For every $n \in \mathbb{N}_0$ and $t > 0$ the operator $I_n(t)$ is an integral operator. Let $p_t(x,y)^n$ be its integral kernel.*

(1) *The map*
$$(0,\infty) \to C^\infty([0,1], C_R^\infty([0,1], M_{2d}(\hat{\Omega}_n\mathcal{A}_\infty))), \ t \mapsto (y \mapsto p_t(\cdot,y)^n)$$
is smooth.

(2) $p_t(x,y)^n = (-1)^n(p_t(y,x)^n)^*$.

(3) *For every $l,m,n \in \mathbb{N}_0$ there are $C,\omega > 0$ such that for $t > 0$ and $x,y \in [0,1]$*
$$|\partial_x^l \partial_y^m p_t(x,y)^n| \leq C(1 + t^{-\frac{l+m+1}{2}})e^{-\omega t}.$$

PROOF. In degree $n = 0$ the assertions hold by Prop. 3.3.4 and Cor. 3.3.6. Let $k_t(x,y)$ be the integral kernel of $e^{-tD_I^2}$.

For $f \in C_R^\infty([0,1], (\hat{\Omega}_{\leq\mu}\mathcal{A}_i)^{2d})$

$$(I_n(t)f)(x) = \int_0^1 \int_0^1 \int_0^1 (1-s)^{n-1} k_{st}(x,r) R(r) p_{(1-s)t}(r,y)^{n-1} f(y) \, dy\, dr\, ds$$

by induction and by the recursion formula above.

We can interchange the integration over r and y.

For the proof of the existence of the integral kernel and of (1) it suffices to show that the map

$$(s,t) \mapsto \left(y \mapsto \int_0^1 (1-s)^{n-1} k_{st}(\cdot,r) R(r) p_{(1-s)t}(r,y)^{n-1} dr\right)$$

is a smooth map from $[0,1] \times (0,\infty)$ to $C^\infty([0,1], C_R^\infty[0,1], M_{2d}(\hat{\Omega}_n\mathcal{A}_i)))$:

For $s \geq \frac{1}{2}$ this follows from the fact that by induction the map

$$(s,t) \mapsto \left(y \mapsto R p_{(1-s)t}(\cdot,y)^{n-1}\right)$$

is a smooth map from $[\frac{1}{2},1] \times (0,\infty)$ to $C^\infty([0,1], C_R^\infty([0,1], M_{2d}(\hat{\Omega}_n\mathcal{A}_i)))$. Furthermore the family $e^{-stD_I^2}$ is uniformly bounded on $C_R^l([0,1], M_{2d}(\hat{\Omega}_n\mathcal{A}_i))$ for any $l \in \mathbb{N}_0$ by Prop. 3.3.3 and depends smoothly on s,t.

For $s \leq \frac{1}{2}$ this holds since

$$(s,t) \mapsto \left(x \mapsto R^* k_{st}(x,\cdot)^*\right)$$

is a smooth map with values in $C^\infty([0,1], C_R^\infty([0,1], M_{2d}(\hat{\Omega}_1\mathcal{A}_i)))$ and since the action of the family $I_{n-1}((1-s)t)^*$ on $C_R^m([0,1], M_{2d}(\hat{\Omega}_{\leq\mu}\mathcal{A}_i)))$ depends smoothly on s,t and is uniformly bounded for any $m \in \mathbb{N}_0$.

Assertion (2) holds since $I_n(t)^* = (-1)^n I_n(t)$.

The preceding arguments and the following facts imply the estimate in (3):

By induction there is $C > 0$ such that the norm of $(y \mapsto R p_{(1-s)t}(\cdot,y)^{n-1})$ in $C^m([0,1], C_R^l([0,1], M_{2d}(\hat{\Omega}_n\mathcal{A}_i)))$ is bounded by $C(1 + t^{-\frac{l+m+1}{2}})e^{-\omega t}$ for $0 \leq s \leq \frac{1}{2}$, $t > 0$. Furthermore by Cor. 3.3.6 the norm of $(x \mapsto R^* k_{st}(x,\cdot)^*)$ is bounded in $C^l([0,1], C_R^m([0,1], M_{2d}(\hat{\Omega}_1\mathcal{A}_i))$ by $C(1 + t^{-\frac{l+m+1}{2}})e^{-\omega t}$ for $s > \frac{1}{2}$, $t > 0$. □

The proof of the previous proposition did not use the fact that $D_I^2 + \sigma R$ is the curvature of a superconnection. Hence an analogous argument shows that $e^{-t(D_{I_s}^2 + \sigma R)}$ is an integral operator whose integral kernel can be written as

$$\sum_{n=0}^{\infty} (-1)^{\frac{(n+1)n}{2}} \sigma^n t^n p_t^s(x,y)^n$$

with $(y \mapsto p_t^s(\cdot, y)^n) \in C^\infty([0,1], C_R^\infty([0,1], M_{2d}(\hat{\Omega}_n \mathcal{A}_i)))$ for $R = (P_s, 1 - P_s)$.

LEMMA 4.1.2. *For every $l, m, n \in \mathbb{N}_0$ and $\varepsilon, \delta > 0$ there is $C > 0$ such that for all $x, y \in [0,1]$ with $d(x,y) > \varepsilon$ and all $t > 0$*

$$|\partial_x^l \partial_y^m p_t^s(x,y)^n| \leq C e^{-\frac{d(x,y)^2}{(4+\delta)t}}.$$

PROOF. In degree $n = 0$ the assertion holds by Lemma 3.1.2.

Let $\eta = \varepsilon/4$.

Let $\chi : \mathbb{R} \to [0,1]$ be a smooth function with $\chi(x) = 0$ for $x > \eta$ and $\chi(x) = 1$ for $x \leq \eta/2$.

Let k_t be the integral kernel of $e^{-tD_{I_s}^2}$.

For $l, m \in \mathbb{N}_0$

$$\partial_x^l \partial_y^m p_t^s(x,y)^n$$
$$= \int_0^1 \int_0^1 (1-s)^{n-1} \partial_x^l k_{st}(x,r) R(r) \chi(d(x,r)) \partial_y^m p_{(1-s)t}^s(r,y)^{n-1} dr ds$$
$$+ \int_0^1 \int_0^1 (1-s)^{n-1} \partial_x^l k_{st}(x,r) R(r) (1 - \chi(d(x,r))) \ldots$$
$$\ldots \chi(d(r,y)) \partial_y^m p_{(1-s)t}^s(r,y)^{n-1} dr ds$$
$$+ \int_0^1 \int_0^1 (1-s)^{n-1} \partial_x^l k_{st}(x,r) R(r) (1 - \chi(d(x,r))) \ldots$$
$$\ldots (1 - \chi(d(r,y))) \partial_y^m p_{(1-s)t}^s(r,y)^{n-1} dr ds.$$

We begin by estimating the first term on the right hand side: By induction there is $C > 0$ such that for $x, y \in [0,1]$ with $d(x,y) > \varepsilon$, $0 < s < 1$ and $t > 0$

$$\|R\chi(d(\cdot,x)) \partial_y^m p_{(1-s)t}^s(\cdot,y)^{n-1}\|_{C_R^l} \leq C e^{-\frac{(d(x,y)-\eta)^2}{(4+\delta)t}}$$

in $C_R^l([0,1], M_{2d}(\hat{\Omega}_n \mathcal{A}_i))$.

Since the operator $e^{-stD_{I_s}^2}$ is uniformly bounded on $C_R^l([0,1], M_{2d}(\hat{\Omega}_n \mathcal{A}_i))$, the first term is bounded by $C e^{-\frac{(d(x,y)-\eta)^2}{(4+\delta)t}}$.

An analogous bound exists for the second term: By Lemma 3.1.2 there is $C > 0$ such that for all $x, y \in [0,1]$ with $d(x,y) > \varepsilon$ and $0 < s < 1$ and $t > 0$ in $C_R^m([0,1], M_{2d}(\hat{\Omega}_n \mathcal{A}_i))$

$$\|(\partial_x^l k_{st}(x,\cdot) R)^* (1 - \chi(d(\cdot,x))) \chi(d(\cdot,y))\|_{C_R^m} \leq C e^{-\frac{(d(x,y)-\eta)^2}{(4+\delta)t}}.$$

Furthermore the integral kernel $(y,r) \mapsto (\partial_y^m p_{(1-s)t}^s(r,y)^{n-1})^*$ induces a uniformly bounded family of operators from $C_R^m([0,1], M_{2d}(\hat{\Omega}_{\leq \mu} \mathcal{A}_i))$ to $C_R([0,1], M_{2d}(\hat{\Omega}_{\leq \mu} \mathcal{A}_i))$.

The third term is bounded by $C e^{-\frac{d(x,y)^2}{(4+\delta)t}}$ since by Lemma 3.1.2

$$|\partial_x^l k_{st}(x,r) R(r) (1 - \chi(d(x,r)))| \leq C e^{-\frac{d(x,r)^2}{(4+\delta)st}}$$

and by induction
$$\left|\bigl(1-\chi(d(x,r))\bigr)\partial_y^m p_{(1-s)t}^s(r,y)^{n-1}\right| \le Ce^{-\frac{d(r,y)^2}{(4+\delta)(1-s)t}}$$
for all $x,y,r \in [0,1]$, all $0 < s < 1$ and $t > 0$.

Hence there is $C > 0$ such that for all $x,y \in [0,1]$ and all $t > 0$
$$|\partial_x^l \partial_y^m p_t^s(x,y)^n| \le Ce^{-\frac{(d(x,y)-\varepsilon/2)^2}{(4+\delta)t}}.$$
The assertion follows now from Lemma 3.1.6.

$\square$

As in §3.3.2 we apply Duhamel's principle in order to obtain an estimate for the kernel $p_t(x,y)^n$:

Recall the definitions of ϕ_k, γ_k, $k = 0, 1$, in §3.3.2.

By Lemma 1.4.3 we can find unitaries $U_k \in M_{2d}(\mathcal{A}_\infty)$, $k = 0, 1$, with $U_k I_0 = I_0 U_k$ and $U_k P_k U_k^* = P_s$.

Let $W_n^k(t)$ be the integral operator with integral kernel
$$w_t^k(x,y)^n := U_k^* p_t^s(x,y)^n U_k$$
and denote by $W_n(t)$ the integral operator of
$$w_t(x,y)^n := \gamma_0(x) w_t^0(x,y)^n \phi_0(y) + \gamma_1(x) w_t^1(x,y)^n \phi_1(y).$$
Set $W_0(0) := 1$ and $W_n(0) := 0$ for $n \ge 1$. Then $W_n(t)$ is a strongly continuous family of operators on $L^2([0,1],(\hat{\Omega}_{\le\mu}\mathcal{A}_i)^{2d})$ for all $n \in \mathbb{N}_0$. For $f \in C_R^\infty([0,1],(\hat{\Omega}_{\le\mu}\mathcal{A}_i)^{2d})$ the map $[0,\infty) \to L^2([0,1],(\hat{\Omega}_{\le\mu}\mathcal{A}_i)^{2d})$, $t \mapsto W_n(t)f \in L^2([0,1],(\hat{\Omega}_{\le\mu}\mathcal{A}_i)^{2d})$ is even smooth.

Furthermore for $t > 0$ the range of $W_n(t)$ is in $C_R^\infty([0,1],(\hat{\Omega}_{\le\mu}\mathcal{A}_i)^{2d})$.

Hence Duhamel's principle yields for $f \in C_R^\infty([0,1],\mathcal{A}_i^{2d})$:

$$\left(e^{-tA_I^2} - \sum_{n=0}^\infty \sigma^n (-1)^{\frac{n(n+1)}{2}} t^n W_n(t)\right)f$$
$$= -\int_0^t e^{-sA_I^2}\left(\frac{d}{dt} + A_I^2\right)\sum_{n=0}^\infty \sigma^n(-1)^{\frac{n(n+1)}{2}}(t-s)^n W_n(t-s)f\, ds$$
$$= \int_0^t e^{-sA_I^2} \sum_{k=0,1}\sum_{n=0}^\infty \sigma^n(-1)^{\frac{n(n+1)}{2}}(t-s)^n [\gamma_k, D_I^2]_s W_n^k(t-s)\phi_k f\, ds$$
$$= \sum_{n=0}^\infty \sigma^n \sum_{j=0}^n (-1)^{k(n,j)} \int_0^t s^{n-j}(t-s)^j I_{n-j}(s)\ldots$$
$$\ldots \sum_{k=0,1}(\gamma_k'\partial + \partial\gamma_k')W_j^k(t-s)\phi_k f\, ds$$
with $k(n,j) = \frac{(n-j)(n-j+1)+j(j+1)}{2}$.

It follows that
$$(*) \quad (I_n(t) - W_n(t))f$$
$$= t^{-n}\sum_{j=0}^n (-1)^{j(n-j)} \int_0^t s^{n-j}(t-s)^j I_{n-j}(s) \sum_{k=0,1}(\gamma_k'\partial + \partial\gamma_k')W_j^k(t-s)\phi_k f\, ds.$$

PROPOSITION 4.1.3. *For every $l, m, n \in \mathbb{N}_0$ and every $\delta > 0$ there is $C > 0$ such that*

$$|\partial_x^l \partial_y^m p_t(x,y)^n - \partial_x^l \partial_y^m w_t(x,y)^n| \le Ct \sum_{k=0,1} e^{-\frac{d(y,\operatorname{supp} \gamma_k')^2}{(4+\delta)t}} 1_{\operatorname{supp} \phi_k}(y)$$

for all $t > 0$ and all $x, y \in [0, 1]$.

PROOF. The proof is as for $e^{-tD_I^2}$ (see Prop. 3.3.4).

From the previous lemma it follows that for every $j \in \mathbb{N}_0$ and $k = 0, 1$ there is $C > 0$ such that

$$\|(\gamma_k' \partial + \partial \gamma_k') \partial_y^m (w_{t-s}^k(\cdot, y)^j \phi_k(y))\|_{C_R^l} \le Ce^{-\frac{d(y,\operatorname{supp} \gamma_k')^2}{(4+\delta)t}} 1_{\operatorname{supp} \phi_k}(y)$$

for all $y \in [0, 1]$, all $t > 0$ and $0 < s < t$. From this estimate, from the fact that $I_{n-j}(s)$ is a uniformly bounded family of operators on $C_R^l([0,1], M_{2d}(\hat{\Omega}_{\le n} \mathcal{A}_i))$ and from the equation $(*)$ one deduces as in the proof of Prop. 3.3.4 that

$$\|\partial_y^m p_t(\cdot, y)^n - \partial_y^m w_t(\cdot, y)^n\|_{C_R^l} \le Ce^{-\frac{d(y,\operatorname{supp} \gamma_k')^2}{(4+\delta)t}} 1_{\operatorname{supp} \phi_k}(y)$$

for all $y \in [0, 1]$, all $t > 0$ and $0 < s < t$. $\square$

COROLLARY 4.1.4. *For every $l, m, n \in \mathbb{N}_0$ and $\varepsilon > 0$ there are $c, C > 0$ such that*

$$|\partial_x^l \partial_y^m p_t(x,y)^n| \le Ce^{-\frac{d(x,y)^2}{ct}}$$

for all $x, y \in [0, 1]$ with $d(x, y) > \varepsilon$ and all $t > 0$.

PROOF. Note that the assertion is equivalent to the assertion that for every $l, m, n \in \mathbb{N}_0$ and $\varepsilon > 0$ there are $c, C > 0$ such that

$$|\partial_x^l \partial_y^m p_t(x,y)^n| \le Ce^{-\frac{1}{ct}}$$

for all $x, y \in [0, 1]$ with $d(x, y) > \varepsilon$ and all $t > 0$. The estimate follows for $0 < t < 1$ from the previous proposition and Lemma 4.1.2 since $\operatorname{supp} \gamma_k' \cap \operatorname{supp} \phi_k = \emptyset$. For $t > 1$ it holds by the estimate in Prop. 4.1.1. $\square$

4.1.3. The η-form. In the following $(D_I p_t)(x,y)^n$ denotes the integral kernel of $D_I I_n(t)$.

LEMMA 4.1.5. *For every $n \in \mathbb{N}_0$ the integral*

$$\int_0^\infty t^{\frac{n-1}{2}} \int_0^1 \operatorname{tr}(D_I p_t)(x,x)^n dx dt$$

is well-defined in $\hat{\Omega}_* \mathcal{A}_\infty / \overline{[\hat{\Omega}_* \mathcal{A}_\infty, \hat{\Omega}_* \mathcal{A}_\infty]}_s$.

PROOF. The integral converges in $\hat{\Omega}_{\le \mu} \mathcal{A}_i / \overline{[\hat{\Omega}_{\le \mu} \mathcal{A}_i, \hat{\Omega}_{\le \mu} \mathcal{A}_i]}_s$ for all $n, \mu, i \in \mathbb{N}_0$: For $n = 0$ and $t \to 0$ the convergence follows from Cor. 3.3.8; for $n > 0$ and $t \to 0$ and for $n \in \mathbb{N}_0$ and $t \to \infty$ it follows from Prop. 4.1.1. For the convergence in $\hat{\Omega}_* \mathcal{A}_\infty / \overline{[\hat{\Omega}_* \mathcal{A}_\infty, \hat{\Omega}_* \mathcal{A}_\infty]}_s$ apply Prop. 1.3.5. $\square$

Let $\operatorname{tr}_\sigma(a + \sigma b) := \operatorname{tr} a$ for $a, b \in M_{2d}(\hat{\Omega}_* \mathcal{A}_i)$ and let Tr_σ be the corresponding trace on integral operators.

4.1. THE SUPERCONNECTION A_t^I ASSOCIATED TO D_I

DEFINITION 4.1.6. *The η-FORM OF THE SUPERCONNECTION A_I is*

$$\eta(A_I) := \frac{1}{\sqrt{\pi}} \int_0^\infty t^{-\frac{1}{2}} \text{Tr}_\sigma D_I e^{-(A_t^I)^2} dt \in \hat{\Omega}_* \mathcal{A}_\infty / \overline{[\hat{\Omega}_* \mathcal{A}_\infty, \hat{\Omega}_* \mathcal{A}_\infty]_s} .$$

Since

$$\eta(A_I) = \frac{1}{\sqrt{\pi}} \sum_{n=0}^\infty (-1)^n \int_0^\infty t^{n-\frac{1}{2}} \int_0^1 \text{tr}(D_I p_t)(x,x)^{2n} \, dx dt$$

the η-form is well-defined by the previous lemma.

For $\varepsilon \in]0, \frac{1}{2}]$ let $A_I(\varepsilon) := U_\varepsilon^* \, dU_\varepsilon + \sigma D_I$ be a family of superconnections with $\text{supp}(U_\varepsilon - 1) \subset [0, \varepsilon] \cup [1 - \varepsilon, 1]$.

PROPOSITION 4.1.7. *The limit $\lim_{\varepsilon \to 0} \eta(A_I(\varepsilon))$ exists and does not depend on the choice of U_ε.*

PROOF. Let $R_\varepsilon := [D_I, U_\varepsilon^* \, dU_\varepsilon]$. It suffices to prove that for $y, z \in [0, 1]$ and $s, t > 0$ the limit

$$\lim_{\varepsilon \to 0} \int_0^1 k_s(y, x) R_\varepsilon(x) k_t(x, z) \, dx$$

exists and does not depend on the choice of U_ε.

Let $f(x) := k_s(y, x)$ and $g(x) := k_t(x, z)$.
Then

$$\int_0^1 f(x) R_\varepsilon(x) g(x) \, dx$$

$$= \int_0^1 f(x) [I_0 \partial_x, U_\varepsilon^*(x) \, dU_\varepsilon(x)] g(x) \, dx$$

$$= \int_0^1 \partial_x \bigl(f(x) I_0 U_\varepsilon^*(x) \, dU_\varepsilon(x) g(x)\bigr) \, dx$$

$$\quad - \int_0^1 f'(x) I_0 U_\varepsilon^*(x) \, dU_\varepsilon(x) g(x) \, dx - \int_0^1 f(x) I_0 U_\varepsilon^*(x) \, dU_\varepsilon(x) g'(x) \, dx$$

$$= f(1) I_0 U_\varepsilon^*(1) \, dU_\varepsilon(1) g(1) - f(0) I_0 U_\varepsilon^*(0) \, dU_\varepsilon(0) g(0)$$

$$\quad - \int_0^1 f'(x) I_0 U_\varepsilon^*(x) \, dU_\varepsilon(x) g(x) \, dx - \int_0^1 f(x) I_0 U_\varepsilon^*(x) \, dU_\varepsilon(x) g'(x) \, dx$$

$$= - \int_0^1 f'(x) I_0 U_\varepsilon^*(x) \, dU_\varepsilon(x) g(x) \, dx - \int_0^1 f(x) I_0 U_\varepsilon^*(x) \, dU_\varepsilon(x) g'(x) \, dx .$$

Since for $\varepsilon \to 0$ and $x \in (0, 1)$ the term $U_\varepsilon^*(x) \, dU_\varepsilon(x) g(x)$ resp. $U_\varepsilon^*(x) \, dU_\varepsilon(x) g'(x)$ converges to $d g(x)$ resp. $d g'(x)$, it follows that

$$\lim_{\varepsilon \to 0} \int_0^1 f(x) R_\varepsilon(x) g(x) \, dx = f(0) I_0 \, d g(0) - f(1) I_0 \, d g(1) .$$

□

DEFINITION 4.1.8. *We call*

$$\eta(P_0, P_1) := \lim_{\varepsilon \to 0} \eta(A_I(\varepsilon))$$

the η-FORM ASSOCIATED TO THE PAIR (P_0, P_1).

4.2. The superconnection associated to D_Z

Let (P_0, P_1) be a pair of transverse Lagrangian projections with $P_i \in M_{2d}(\mathcal{A}_\infty)$, $i = 1, 2$, and let D_Z be the operator on $L^2(Z, (\hat{\Omega}_{\leq \mu} \mathcal{A}_i)^{4d})$ with boundary conditions (P_0, P_1) as in §3.4.

Let $U \in C^\infty([0,1], M_{2d}(\mathcal{A}_\infty))$ be as in Prop. 2.2.1 with $U(0) P_0 U(0)^* = P_s$ and $U(1) P_1 U(1)^* = 1 - P_s$. We consider U as a function on Z depending only on the variable x_2 and set
$$\tilde{W} := U \oplus U \in C^\infty(Z, M_{2d}(\mathcal{A}_\infty)) \ .$$
We call $A_Z := \tilde{W}^* \, d\tilde{W} + D_Z$ a superconnection associated to D_Z and $A_t^Z := \tilde{W}^* \, d\tilde{W} + \sqrt{t} D_Z$ the corresponding rescaled superconnection.

The curvature is
$$\begin{aligned} A_Z^2 &= D_Z^2 + [\tilde{W}^* \, d\tilde{W}, D_Z]_s \\ &= D_Z^2 + \tilde{W}^*[d, \tilde{W} c(dx_2) \partial_{x_2} \tilde{W}^*]_s \tilde{W} \\ &= D_Z^2 - c(dx_1) \tilde{W}^* [d, \tilde{W} I \partial_{x_2} \tilde{W}^*]_s \tilde{W} \\ &= D_Z^2 - c(dx_1) \tilde{W}^* \, d(\tilde{W} I \partial_{x_2} \tilde{W}^*) \tilde{W} \\ &= D_Z^2 + c(dx_1)(R \oplus (-R)) \end{aligned}$$
with $R = -U^* \, d(U I_0(\partial U^*)) U$.

Let $\tilde{R} = c(dx_1)(R \oplus (-R))$. Then $\tilde{R} \in C^\infty(Z, M_{4d}(\hat{\Omega}_{\leq \mu} \mathcal{A}_i))$ and $\tilde{R}^* = \tilde{R}$.

We have a holomorphic semigroup $e^{-t A_Z^2}$ on $L^2(Z, (\hat{\Omega}_{\leq \mu} \mathcal{A}_i)^{4d})$.

From §3.4 recall that $e^{-t D_Z^2} = e^{-t \tilde{D}_I^2} e^{-t \Delta_\mathbb{R}}$.

By Prop. 4.1.1 the operator
$$I_n(t) = \int_{\Delta^n} e^{-u_0 t D_I^2} R e^{-u_1 t D_I^2} R \ldots e^{-u_n t D_I^2} \, du_0 \ldots du_n$$
is an integral operator. Its integral kernel is denoted by $p_t^I(x,y)^n$.

Since $e^{-t \Delta_\mathbb{R}}$ commutes with $\tilde{R}$ and $c(dx_1)$ commutes with $e^{-t D_Z^2}$ and $R \oplus (-R)$, we obtain from Volterra development:
$$\begin{aligned} e^{-t A_Z^2} &= \sum_{n=0}^\infty (-1)^n t^n \int_{\Delta^n} e^{-u_0 t D_Z^2} \tilde{R} e^{-u_1 t D_Z^2} \tilde{R} \ldots e^{-u_n t D_Z^2} \, du_0 \ldots du_n \\ &= \sum_{n=0}^\infty (-1)^n t^n e^{-t \Delta_\mathbb{R}} \int_{\Delta^n} e^{-u_0 t \tilde{D}_I^2} \tilde{R} e^{-u_1 t \tilde{D}_I^2} \tilde{R} \ldots e^{-u_n t \tilde{D}_I^2} \, du_0 \ldots du_n \\ &= \sum_{n=0}^\infty (-1)^n t^n c(dx_1)^n e^{-t \Delta_\mathbb{R}} \left(I_n(t) \oplus (-1)^n I_n(t) \right) \ . \end{aligned}$$

We define
$$p_t^Z(x,y)^n := c(dx_1)^n \frac{1}{\sqrt{4\pi t}} e^{-\frac{(x_1-y_1)^2}{4t}} \left(p_t^I(x_2,y_2)^n \oplus (-1)^n p_t^I(x_2,y_2)^n \right) \ .$$
The integral kernel of $e^{-t A_Z^2}$ is
$$\sum_{n=0}^\infty (-1)^n t^n p_t^Z(x,y)^n$$

and the integral kernel of $e^{-(A_t^Z)^2}$

$$\sum_{n=0}^{\infty}(-1)^n t^{\frac{n}{2}} p_t^Z(x,y)^n .$$

Note that for all multi-indices $\alpha, \beta \in \mathbb{N}_0^2$

$$\operatorname{tr}_s \partial_x^\alpha \partial_y^\beta p_t^Z(x,y)^n = 0 .$$

Furthermore

$$(D_Z)_x p_t^Z(x,y)^n = c(dx_1)(\partial_{x_1} + I\partial_{x_2})p_t^Z(x,y)^n .$$

Since $\operatorname{tr}_s c(dx_1)\partial_{x_1} p_t^Z(x,y)^n$ vanishes for $x_1 = y_1$ and $\operatorname{tr}_s c(dx_1) I \partial_{x_2} p_t^Z(x,y)^n$ vanishes for all $x, y \in Z$,

$$\operatorname{tr}_s (D_Z)_x p_t^Z(x,y)^n = 0$$

if $x = y$.

Furthermore the integral kernel $p_t^Z(x,y)^n$ satisfies the following Gaussian estimate:

LEMMA 4.2.1. *Let $\alpha, \beta \in \mathbb{N}_0^2$. For every $\varepsilon > 0$ there are $c, C > 0$ such that for all $x, y \in Z$ with $d(x,y) > \varepsilon$ and all $t > 0$*

$$|\partial_x^\alpha \partial_y^\beta p_t^Z(x,y)^n| \le C e^{-\frac{d(x,y)^2}{ct}} .$$

PROOF. The assertion follows from Prop. 4.1.1 and Cor. 4.1.4. The arguments are as in the proof of Lemma 3.4.4. □

4.3. The superconnection $A(\rho)_t$ associated to $D(\rho)$

4.3.1. The family $e^{-A(\rho)_t^2}$. In §3.5.1 we fixed $r_0, b_0 > 0$ such that

$$\operatorname{supp} k_K \cap \big((F(r_0, b_0) \times M) \cup (M \times F(r_0, b_0))\big) = \emptyset .$$

Let $W \in C^\infty(M, \operatorname{End}^+ E \otimes \mathcal{A}_\infty)$ be as in §2.1.2 such that W is parallel on $\{x \in M \mid d(x, \partial M) > b_0\}$. Then $[W, K]_s = 0$.

We define a superconnection associated to $D(\rho)$ on the $\mathbb{Z}/2$-graded $\mathcal{A}_i$-module $L^2(M, E \otimes \mathcal{A}_i)$ by

$$A(\rho) := W^* \operatorname{d} W + D(\rho) .$$

The corresponding rescaled superconnection is

$$A(\rho)_t := W^* \operatorname{d} W + \sqrt{t} D(\rho) .$$

The curvature of $A(\rho)$ is

$$\begin{aligned} A(\rho)^2 &= W^* \operatorname{d}^2 W + [D(\rho), W^* \operatorname{d} W]_s + D(\rho)^2 \\ &= D(\rho)^2 + W^*[\operatorname{d}, Wc(dW^*)]_s W \\ &=: D(\rho)^2 + \mathcal{R} . \end{aligned}$$

We used Prop. 2.1.1 and $[W, K]_s = 0$.

Then $\mathcal{R} \in C^\infty(M, \operatorname{End} E \otimes \hat{\Omega}_{\le 1} \mathcal{A}_i)$ with $\mathcal{R}^* = \mathcal{R}$. In the flat region

$$\begin{aligned} \mathcal{R}|_F &= W^*[\operatorname{d}, Wc(e_2)\partial_{e_2} W^*]_s W \\ &= -c(e_1) W^* \operatorname{d}(WI\partial_{e_2} W^*) W . \end{aligned}$$

Furthermore $\mathcal{R}$ vanishes on $\{x \in M \mid d(x, \partial M) > b_0\}$.

For every $k \in \mathbb{Z}/6$ the restriction of $\mathcal{R}$ to $\mathcal{U}_k$ is of the form $c(e_1)(R \oplus (-R))$ with $R \in C^\infty(\mathcal{U}_k, M_{2d}(\hat{\Omega}_{\le 1} \mathcal{A}_i))$ and R is independent of the variable x_2^k.

The rescaled curvature is

$$A(\rho)_t^2 = tD(\rho)^2 + \sqrt{t}\mathcal{R} \ .$$

As expected, $A(\rho)^2$ and $A(\rho)_t^2$ are right $\hat{\Omega}_{\leq\mu}\mathcal{A}_i$-module homomorphisms.

Since $A(\rho)^2$ is a bounded perturbation of $D(\rho)^2$, it generates a holomorphic semigroup $e^{-tA(\rho)^2}$ on $L^2(M, E \otimes \hat{\Omega}_{\leq\mu}\mathcal{A}_i)$.

In the following we assume that $t \geq 0$.

By Volterra development

$$\begin{aligned} e^{-tA(\rho)^2} &= \sum_{n=0}^{\infty}(-1)^n t^n \int_{\Delta^n} e^{-u_0 tD(\rho)^2}\mathcal{R}e^{-u_1 tD(\rho)^2}\mathcal{R}\ldots e^{-u_n tD(\rho)^2}\,du_0\ldots du_n \\ &=: \sum_{n=0}^{\infty}(-1)^n t^n I_n(\rho, t) \ . \end{aligned}$$

It follows that

$$e^{-A(\rho)_t^2} = \sum_{n=0}^{\infty}(-1)^n t^{n/2} I_n(\rho, t) \ .$$

For $\rho \neq 0$ the family $t \mapsto I_n(\rho, t)$ is uniformly bounded on $L^2(M, E \otimes \hat{\Omega}_{\leq\mu}\mathcal{A}_i)$. By Cor. 3.5.13 it acts as a strongly continuous family of operators on $C_R^m(M, E \otimes \hat{\Omega}_{\leq\mu}\mathcal{A}_i)$ and there are $C, l > 0$ such that the action is bounded by $C(1+t)^l$.

4.3.2. The integral kernel of $e^{-A(\rho)_t^2}$. In this section we prove that $I_n(\rho, t)$ is an integral operator for $t > 0$. As usual we construct an approximation of the family $I_n(\rho, t)$ by a family of integral operators and compare it with $I_n(\rho, t)$ by Duhamel's principle.

Let $\mathcal{U}(r_0, b_0) = \{\mathcal{U}_k\}_{k \in J}$, $\{\phi_k\}_{k \in J}$ and $\{\gamma_k\}_{k \in J}$ be as in §3.5.1.

For $k \in \mathbb{Z}/6$ the function $W|_{\mathcal{U}_k} : \mathcal{U}_k \to M_{4d}(\mathcal{A}_\infty)$ does not depend on the coordinate x_1^k. We extend it to a section $\tilde{W}_k : Z_k \to M_{4d}(\mathcal{A}_\infty)$ independent of x_1^k and define the superconnection $A_{Z_k} := \tilde{W}_k^* \,d\, \tilde{W}_k + D_{Z_k}$, which coincides on $\mathcal{U}_k$ with the superconnection $A(\rho)$.

For $k \in \mathbb{Z}/6$ and $n \in \mathbb{N}_0$ let $w(\rho)_t^k(x,y)^n$ be the restriction of $p_t^{Z_k}(x,y)^n$ to $\mathcal{U}_k \times \mathcal{U}_k$.

Let furthermore $w(\rho)_t^{\clubsuit}(x,y)^0$ be the restriction of the integral kernel of $e^{-tD_N(\rho)^2}$ to $\mathcal{U}_{\clubsuit} \times \mathcal{U}_{\clubsuit}$ and set $w(\rho)_t^{\clubsuit}(x,y)^n = 0$ for $n > 0$.

The reason for this is that $A(\rho)^2$ equals $D(\rho)^2$ on $\mathcal{U}_{\clubsuit}$.

We extend $w(\rho)_t^k(x,y)^n$ by zero to $M \times M$ and set

$$w(\rho)_t(x,y)^n := \sum_{k \in J} \gamma_k(x) w(\rho)_t^k(x,y)^n \phi_k(y) \ .$$

Write $W_n(\rho, t)$ for the corresponding integral operator. It is a bounded operator on $L^2(M, E \otimes \hat{\Omega}_{\leq\mu}\mathcal{A}_i)$ and on $C_R^m(M, E \otimes \hat{\Omega}_{\leq\mu}\mathcal{A}_i)$.

Set $W_0(\rho, 0) = 1$ and $W_n(\rho, 0) = 0$ for $n > 0$.

Then for $f \in L^2(M, E \otimes \hat{\Omega}_{\leq\mu}\mathcal{A}_i)$ the family $W_n(\rho, t)f \in L^2(M, E \otimes \hat{\Omega}_{\leq\mu}\mathcal{A}_i)$ depends continuously on t for all $t \in [0, \infty)$, and for $f \in C_{Rc}^\infty(M, E \otimes \mathcal{A}_i)$ even smoothly.

4.3. THE SUPERCONNECTION $A(\rho)_t$ ASSOCIATED TO $D(\rho)$

For $f \in C^\infty_{Rc}(M, E \otimes \mathcal{A}_i)$ Duhamel's principle yields:

$$\left(e^{-tA(\rho)^2} - \sum_{n=0}^\infty (-1)^n t^n W_n(\rho, t)\right) f$$

$$= -\int_0^t e^{-sA(\rho)^2} \left(\frac{d}{dt} + A(\rho)^2\right) \sum_{k \in J} \sum_{n=0}^\infty (-1)^n (t-s)^n \gamma_k W_n^k(\rho, t-s) \phi_k f \, ds$$

$$= \int_0^t e^{-sA(\rho)^2} \sum_{k \in J} [\gamma_k, D(\rho)^2]_s \sum_{n=0}^\infty (-1)^n (t-s)^n W_n^k(\rho, t-s) \phi_k f \, ds$$

$$= \sum_{n=0}^\infty (-1)^n \int_0^t \sum_{m=0}^n s^{n-m}(t-s)^m I_{n-m}(\rho, s) \ldots$$

$$\ldots \sum_{k \in J} [\gamma_k, D(\rho)^2]_s W_m^k(\rho, t-s) \phi_k f \, ds \, .$$

Hence

$$\left(I_n(\rho, t) - W_n(\rho, t)\right) f$$

$$= -t^{-n} \int_0^t \sum_{m=0}^n s^{n-m}(t-s)^m I_{n-m}(\rho, s) \sum_{k \in J} [\gamma_k, D(\rho)^2]_s W_m^k(\rho, t-s) \phi_k f \, ds \, .$$

In the following proposition $|\cdot|$ denotes the norm on the fibers of $(E \boxtimes E^*) \otimes \hat{\Omega}_{\leq \mu} \mathcal{A}_i$.

PROPOSITION 4.3.1. *The operator $I_n(\rho, t)$ is an integral operator for $t > 0$. Let $p(\rho)_t(x, y)^n$ be its integral kernel.*

(1) *The map $(0, \infty) \to C^\infty(M \times M, (E \boxtimes E^*) \otimes \hat{\Omega}_n \mathcal{A}_\infty), t \mapsto p(\rho)_t^n$ is smooth.*
(2) $p(\rho)_t(x, y)^n = (p(\rho)_t(y, x)^n)^*$.
(3) *For every $T > 0$ there are $C, c > 0$ such that*

$$|p(\rho)_t(x, y)^n - w(\rho)_t(x, y)^n| \leq Ct \left(|\rho| 1_{\mathcal{U}_\clubsuit}(y) + \sum_{k \in J} e^{-\frac{d(y, \text{supp } d\gamma_k)^2}{ct}} 1_{\text{supp } \phi_k}(y)\right)$$

for all $0 < t < T$, $\rho \in [-1, 1]$ and all $x, y \in M$.

(4) *Let $\rho \neq 0$. Then there are $C, c > 0$ and $j \in \mathbb{N}$ such that*

$$|p(\rho)_t(x, y)^n - w(\rho)_t(x, y)^n| \leq Ct(1+t)^j e^{-\frac{d(y, \mathcal{U}_\clubsuit)^2}{ct}}$$

for all $t > 0$ and all $x, y \in M$.

Statements analogous to (3), (4) hold for the partial derivatives of $p(\rho)_t(x, y)^n$ in x of y with respect to unit vector fields on M.

PROOF. The proof follows the proof of Prop. 3.5.11.

In order to show the existence of the integral kernel and (1) we need only investigate $I_n(\rho, t) - W_n(\rho, t)$.

For $f \in C^\infty_{Rc}(M, E \otimes \mathcal{A}_i)$

$$\left(I_n(\rho, t) - W_n(\rho, t)\right) f$$

$$= t^{-n} \int_0^t \int_M \sum_{m=0}^n s^{n-m}(t-s)^m I_{n-m}(\rho, s) \sum_{k \in J} [c(d\gamma_k), D]_s w(\rho)_{t-s}^k(\cdot, y)^m \phi_k(y) f(y) \, dy \, ds \, .$$

For $\rho \in \mathbb{R}$ and $t > 0$ the family $I_{n-m}(\rho, s)$ is uniformly bounded on $C_R^\nu(M, E \otimes \hat{\Omega}_{\leq \mu}\mathcal{A}_i)$ in $s < t$.

The function
$$\tau \mapsto \left(y \mapsto \sum_{k \in J} [c(d\gamma_k), D]_s w(\rho)_\tau^k(\cdot, y)^m \phi_k(y)\right)$$
is smooth from $(0, \infty)$ to $C^l(M, C_{Rc}^\nu(M, E \otimes \hat{\Omega}_{\leq m+1}\mathcal{A}_i) \otimes E^*)$ for any $l, \nu \in \mathbb{N}_0$.

If $k \in \mathbb{Z}/6$, then by Lemma 4.2.1 there are $c, C > 0$ such that for all $y \in M$ and $0 < \tau$
$$\|[c(d\gamma_k), D]_s w_\tau^k(\cdot, y)^m \phi_k(y)\|_{C^\nu} \leq C e^{-\frac{d(y, \text{supp } d\gamma_k)^2}{c\tau}} 1_{\text{supp } \phi_k}(y) \ .$$

Furthermore $w(\rho)_\tau^\clubsuit(x, y)^m = 0$ for $m > 0$. The kernel $w(\rho)_\tau^\clubsuit(x, y)^0$ is equal to the kernel $e(\rho)_\tau^\clubsuit(x, y)$ in the proof of Prop. 3.5.11 and was estimated there. It follows that for $T, \delta > 0$ there is $C > 0$ such that
$$\|[c(d\gamma_k), D]_s w(\rho)_\tau^\clubsuit(\cdot, y)^0 \phi_\clubsuit(y)\|_{C^\nu} \leq C\left(e^{-\frac{d(y, \text{supp } d\gamma_\clubsuit)^2}{(4+\delta)\tau}} + \tau|\rho|\right) 1_{\text{supp } \phi_\clubsuit}(y)$$
for $y \in M$, $\rho \in [-1, 1]$ and $0 < \tau < T$.

Analogous estimates hold for the derivatives in y and also in τ by the heat equation. This shows that the function above extends smoothly to $\tau = 0$. The existence and (1) and (3) follow now by the usual arguments.

Assertion (2) follows from $I_n(\rho, t)^* = I_n(\rho, t)$.

(4) follows from the estimates by taking into account that for every $\rho \neq 0$ there is $C > 0$ and $j \in \mathbb{N}$ such that the norm of $I_n(\rho, t)$ on $C_R^l(M, E \otimes \hat{\Omega}_{\leq \mu}\mathcal{A}_i)$ is bounded by $C(1+t)^j$ for all $t > 0$. $\square$

We deduce the following estimates for later use:

COROLLARY 4.3.2. *Let $\rho \neq 0$. For every $\varepsilon > 0$ and $m, n \in \mathbb{N}_0$ there are $c, C > 0$ and $j \in \mathbb{N}$ such that for $t > 0$ and $x, y \in M$ with $d(x, y) > \varepsilon$*
$$|D(\rho)_x^m p(\rho)_t(x, y)^n| \leq C(1+t)^j \left(e^{-\frac{d(x,y)^2}{ct}} + e^{-\frac{d(y, \mathcal{U}_\clubsuit)^2}{ct}}\right).$$

PROOF. This follows from the proposition and Lemma 4.2.1. $\square$

COROLLARY 4.3.3. *Let $k \in \mathbb{Z}/6$.*
(1) *For every $T > 0$ and $m, n \in \mathbb{N}_0$ there are $c, C > 0$ such that for all $x, y \in \mathcal{U}_k$, for $0 < t < T$ and $\rho \in [-1, 1]$*
$$|D(\rho)_x^m p(\rho)_t(x, y)^n - (D_{Z_k})_x^m p_t^{Z_k}(x, y)^n| \leq C e^{-\frac{d(y, \mathcal{U}_\clubsuit)}{ct}} \ .$$
(2) *For every $\rho \neq 0$ and $n \in \mathbb{N}_0$ there are $c, C > 0$ and $j \in \mathbb{N}$ such that for all $x, y \in \mathcal{U}_k$ and $t > 0$*
$$|D(\rho)_x^m p(\rho)_t(x, y)^n - (D_{Z_k})_x^m p_t^{Z_k}(x, y)^n| \leq C(1+t)^j e^{-\frac{d(y, \mathcal{U}_\clubsuit)}{ct}} \ .$$

4.4. The index theorem and its proof

4.4.1. The generalized supertrace. In the following tr_s denotes the supertrace on the fibers of $(E \otimes E^*) \otimes \hat{\Omega}_{\leq \mu}\mathcal{A}_i$ and Tr_s the corresponding supertrace for trace class operators (see §1.3.2).

Let $\chi : \mathbb{R} \to [0, 1]$ be a smooth function with $\chi(x) = 1$ for $x \leq 0$ and $\chi(x) = 0$ for $x \geq 1$. Let $\phi_r : M \to [0, 1]$, $\phi_r(x) := \chi(d(M_r, x))$ for $r > 0$.

DEFINITION 4.4.1. Let K be a bounded operator on $L^2(M, E \otimes \hat{\Omega}_{\leq \mu} \mathcal{A}_i)$ such that $\phi_r K \phi_r$ is a trace class operator for all $r \geq 0$. Then we define the generalized supertrace
$$\mathrm{Tr}_s K := \lim_{r \to \infty} \mathrm{Tr}_s(\phi_r K \phi_r)$$
if the limit exists.

For trace class operators the generalized supertrace coincides with the usual one. Note that in general it does not vanish on supercommutators.

PROPOSITION 4.4.2. For $\rho \in \mathbb{R}$ and $t > 0$ the generalized supertraces
$$\mathrm{Tr}_s e^{-A(\rho)_t^2}$$
and
$$\mathrm{Tr}_s D(\rho) e^{-A(\rho)_t^2}$$
exist.

PROOF. We show the assertion for $e^{-A(\rho)_t^2}$, the proof for $D(\rho) e^{-A(\rho)_t^2}$ is analogous.

The operator $\phi_r e^{-A(\rho)_t^2} \phi_r$ is trace class for $t > 0$ since
$$\phi_r e^{-A(\rho)_t^2} \phi_r = (\phi_r e^{-A(\rho)_t^2/2})(e^{-A(\rho)_t^2/2} \phi_r)$$
is a product of Hilbert-Schmidt operators. The operators $\phi_r I_n(\rho, t) \phi_r$ are also trace class for $t > 0$.

We show that $\mathrm{tr}_s p(\rho)_t(x, x)^n$ is in $L^1(M, \hat{\Omega}_{\leq \mu} \mathcal{A}_i / \overline{[\hat{\Omega}_{\leq \mu} \mathcal{A}_i, \hat{\Omega}_{\leq \mu} \mathcal{A}_i]_s})$.
By Cor. 4.3.3 there are $c, C > 0$ such that
$$|p(\rho)_t(x, x)^n - p_t^{Z_k}(x, x)^n| \leq C e^{-cd(x, \mathcal{U}_*)^2}$$
for all $x \in \mathcal{U}_k$.
Since $\mathrm{tr}_s p_t^{Z_k}(x, x)^n = 0$ by §4.2,
$$|\mathrm{tr}_s p(\rho)_t(x, x)^n| \leq C e^{-cd(x, \mathcal{U}_*)^2}.$$
Now the assertion follows. □

4.4.2. The limit of $\mathrm{Tr}_s e^{-A(\rho)_t^2}$ for $t \to \infty$. This section is devoted to the proof of

THEOREM 4.4.3. Let $\rho \neq 0$.
Let P_0 be the projection onto the kernel of $D(\rho)$. Then for $T > 0$ there is $C > 0$ such that for all $t > T$
$$\left| \mathrm{Tr}_s e^{-A(\rho)_t^2} - \sum_{n=0}^{\infty} (-1)^n \frac{1}{n!} \mathrm{Tr}_s (P_0 W \, dW^* P_0)^{2n} \right| \leq C t^{-\frac{1}{2}}$$
and
$$|\mathrm{Tr}_s D(\rho) e^{-A(\rho)_t^2}| \leq C t^{-1}$$
in $\hat{\Omega}_{\leq \mu} \mathcal{A}_i / \overline{[\hat{\Omega}_{\leq \mu} \mathcal{A}_i, \hat{\Omega}_{\leq \mu} \mathcal{A}_i]_s}$.

Note that $(P_0 W \, dW^* P_0)^{2n} = W(W^* P_0 W)(d(W^* P_0 W))^{2n} W^*$ by Lemma 5.3.4, hence $(P_0 W \, dW^* P_0)^{2n}$ is a trace class operator on $L^2(M, E \otimes \hat{\Omega}_{\leq \mu} \mathcal{A}_i)$.

The proof is subdivided into some lemmata.

Throughout the section $\rho \neq 0$ is fixed.

In the following $|\cdot|$ denotes the norm on $\hat{\Omega}_{\leq\mu}\mathcal{A}_i/[\overline{\hat{\Omega}_{\leq\mu}\mathcal{A}_i,\hat{\Omega}_{\leq\mu}\mathcal{A}_i}]_s$ resp. the fiberwise norm of $(E\boxtimes E^*)\otimes\hat{\Omega}_{\leq\mu}\mathcal{A}_i$ (depending on the context), and $\|\cdot\|$ denotes the operator norm of $B(L^2(M, E\otimes\hat{\Omega}_{\leq\mu}\mathcal{A}_i))$. Furthermore we make use of the Hilbert-Schmidt norm $\|\ \|_{HS}$ and the trace class norm $\|\ \|_{TR}$, which are defined in §5.2.4 and §5.2.5.

LEMMA 4.4.4. *Let $\nu = 0,1$. For any $T > 0$ there are $\varepsilon, C > 0$ such that for all $t > T$*
$$|\mathrm{Tr}_s D(\rho)^\nu e^{-A(\rho)_t^2} - \mathrm{Tr}_s \phi_t^2 D(\rho)^\nu e^{-A(\rho)_t^2}| \leq Ce^{-\varepsilon t}.$$

PROOF. We prove the case $\nu = 0$, the case $\nu = 1$ can be proved analogously. By Cor. 4.3.3 there are $c, C, r > 0$ such that
$$|p(\rho)_t(x,x)^n - p_t^{Z_k}(x,x)^n| \leq C(1+t)^j e^{-\frac{d(x,M_r)^2}{ct}}$$
for $t > 0$ and $x \in \mathcal{U}_k$ with $k \in \mathbb{Z}/6$.

Hence for $T > 0$ there are $c, C > 0$ such that for all $x \in M$ and $t > T$
$$|\mathrm{tr}_s p(\rho)_t(x,x)^n| \leq Ce^{-\frac{d(x,M_r)^2}{ct}}$$
and thus there are $C, \varepsilon > 0$ such that for all $t > r$
$$\begin{aligned}|\mathrm{Tr}_s e^{-A(\rho)_t^2} - \mathrm{Tr}_s \phi_t^2 e^{-A(\rho)_t^2}| &= |\mathrm{Tr}_s(1-\phi_t^2)e^{-A(\rho)_t^2}| \\ &\leq C\int_t^\infty e^{-\frac{(r'-r)^2}{ct}}dr' \\ &\leq Ce^{-\varepsilon t}.\end{aligned}$$
□

Let $P_1 := 1 - P_0$.

LEMMA 4.4.5. *Let $\nu = 0,1$ and $k \in \mathbb{N}_0$. Then the integral*
$$\int_{\Delta^k} \phi_t D(\rho)^\nu P_1 e^{-u_0 t D(\rho)^2}\mathcal{R}P_1 e^{-u_1 t D(\rho)^2}\mathcal{R}\ldots\mathcal{R}P_1 e^{-u_k t D(\rho)^2}\phi_t\, du_0\ldots du_k$$
converges in the space of trace class operators and for $T > 0$ there are $C, \varepsilon > 0$ such that for all $t > T$ the norm
$$\|\int_{\Delta^k}\left(\phi_t D(\rho)^\nu P_1 e^{-u_0 t D(\rho)^2}\mathcal{R}P_1 e^{-u_1 t D(\rho)^2}\mathcal{R}\ldots\mathcal{R}P_1 e^{-u_k t D(\rho)^2}\phi_t\right) du_0\ldots du_k\|_{TR}$$
is bounded by $Ce^{-\varepsilon t}$.

PROOF. Note that for any $(u_0,\ldots u_k) \in \Delta_k$ we can find $i \in \{0,1,\ldots,k\}$ such that $u_i \geq \frac{1}{k+1}$.

We begin by showing that for any $T > 0$ there are $C, \varepsilon > 0$ such that for $\frac{1}{k+1} \leq u_i \leq 1$, for $0 < u_0,\ldots,u_{i-1} \leq 1$ and for $t > T$ the family
$$\phi_t D(\rho)^\nu P_1 e^{-u_0 t D(\rho)^2}\mathcal{R}P_1 e^{-u_1 t D(\rho)^2}\mathcal{R}\ldots\mathcal{R}P_1 e^{-\frac{u_i}{2} t D(\rho)^2}$$
is a family of Hilbert-Schmidt operators with Hilbert-Schmidt norm bounded by $Cu_0^{-\frac{\nu}{2}}e^{-\varepsilon t}$. If not specified the estimates in the following hold for $\frac{1}{k+1} \leq u_i \leq 1$, for $0 < u_0,\ldots,u_{i-1} \leq 1$ and for $t > T$.

4.4. THE INDEX THEOREM AND ITS PROOF

We have that

$$\phi_t D(\rho)^\nu P_1 e^{-u_0 t D(\rho)^2} \mathcal{R} P_1 e^{-u_1 t D(\rho)^2} \mathcal{R} \ldots \mathcal{R} P_1 e^{-\frac{u_i}{2} t D(\rho)^2}$$
$$= \phi_t D(\rho)^\nu P_1 e^{-u_0 t D(\rho)^2} \mathcal{R} P_1 e^{-u_1 t D(\rho)^2} \mathcal{R} \ldots \mathcal{R} P_1 e^{-\frac{u_i}{2} t D(\rho)^2} \phi_{(t+6)}$$
$$+ \phi_t D(\rho)^\nu P_1 e^{-u_0 t D(\rho)^2} \mathcal{R} \ldots P_1 e^{-u_{i-1} t D(\rho)^2} \mathcal{R} \phi_{(t+6)} P_1 e^{-\frac{u_i}{2} t D(\rho)^2} (1 - \phi_{t+6})$$
$$+ \phi_t D(\rho)^\nu P_1 e^{-u_0 t D(\rho)^2} \mathcal{R} \ldots P_1 e^{-u_{i-1} t D(\rho)^2} \mathcal{R} (1 - \phi_{(t+6)}) P_1 e^{-\frac{u_i}{2} t D(\rho)^2} (1 - \phi_{(t+6)}) \ .$$

Consider the first term on the right hand side. By Cor. 3.5.15 the Hilbert-Schmidt norm of $e^{-\frac{u_i}{4} t D(\rho)^2} \phi_{(t+6)}$ is bounded by Ct.

Furthermore by Prop. 3.5.9 and Cor. 3.5.12 there are $\varepsilon, C > 0$ such that the operator norm of

$$\phi_t D(\rho)^\nu P_1 e^{-u_0 t D(\rho)^2} \mathcal{R} P_1 e^{-u_1 t D(\rho)^2} \mathcal{R} \ldots \mathcal{R} P_1 e^{-\frac{u_i}{4} t D(\rho)^2}$$

is bounded by $C u_0^{-\frac{\nu}{2}} e^{-\varepsilon t}$.

Hence (see Prop. 5.2.13) the first term is a family of Hilbert-Schmidt operators with Hilbert-Schmidt norm bounded by $C u_0^{-\frac{\nu}{2}} e^{-\varepsilon t}$.

In the second term the factor

$$\phi_{(t+6)} P_1 e^{-\frac{u_i}{2} t D(\rho)^2} (1 - \phi_{(t+6)}) = (\phi_{(t+6)} e^{-\frac{u_i}{4} t D(\rho)^2}) P_1 e^{-\frac{u_i}{4} t D(\rho)^2} (1 - \phi_{(t+6)})$$

is a Hilbert-Schmidt operator bounded by $C e^{-\varepsilon t}$ for some $C, \varepsilon > 0$. Hence the second term is bounded in the Hilbert-Schmidt norm by $C u_0^{-\frac{\nu}{2}} e^{-\varepsilon t}$.

The estimate of the third term requires more effort. We prove by induction on $j \in \mathbb{N}$ that there is $C > 0$ such that

$$\phi_t D(\rho)^\nu P_1 e^{-u_0 t D(\rho)^2} \mathcal{R} P_1 e^{-u_1 t D(\rho)^2} \mathcal{R} \ldots P_1 e^{-u_j t D(\rho)^2} (1 - \phi_{(t+6)})$$

is a Hilbert-Schmidt operator with Hilbert-Schmidt norm bounded by $C u_0^{-\frac{\nu}{2}} (1+t)$ for $t > T$ and $0 < u_0, \ldots, u_j \le 1$. Then it follows that the third term is uniformly bounded by $C e^{-\varepsilon t}$ for some $C, \varepsilon > 0$ since $P_1 e^{-\frac{u_i}{2} t D(\rho)^2}$ is exponentially decaying for $t \to \infty$ by Prop. 3.5.9.

For $j = 0$ the assertion follows from Cor. 3.5.15 by

$$\phi_t D(\rho)^\nu P_1 e^{-u_0 t D(\rho)^2} (1 - \phi_{(t+6)})$$
$$= \phi_t D(\rho)^\nu e^{-u_0 t D(\rho)^2} (1 - \phi_{(t+6)}) - \phi_t D(\rho)^\nu P_0 (1 - \phi_{(t+6)}) \ .$$

Now assume the assertion is true for $j - 1$. We have that

$$\phi_t P_1 D(\rho)^\nu e^{-u_0 t D(\rho)^2} \mathcal{R} P_1 \ldots \mathcal{R} P_1 e^{-u_j t D(\rho)^2} (1 - \phi_{(t+6)})$$
$$= \phi_t P_1 D(\rho)^\nu e^{-u_0 t D(\rho)^2} P_1 \mathcal{R} P_1 \ldots$$
$$\ldots P_1 e^{-u_{j-1} t D(\rho)^2} P_1 \mathcal{R} \phi_{(t+3)} P_1 e^{-u_j t D(\rho)^2} (1 - \phi_{(t+6)})$$
$$+ \phi_t P_1 D(\rho)^\nu e^{-u_0 t D(\rho)^2} P_1 \ldots$$
$$\ldots P_1 e^{-u_{j-1} t D(\rho)^2} P_1 (1 - \phi_{(t+3)}) \mathcal{R} P_1 e^{-u_j t D(\rho)^2} (1 - \phi_{(t+6)}) \ .$$

Both terms on the right hand side are bounded in the Hilbert-Schmidt norm by $C u_0^{-\nu/2}(1 + t)$ for all $t > T$ and $0 < u_0, \ldots, u_j \le 1$: the first term since $\phi_{(t+3)} P_1 e^{-u_j t D(\rho)^2} (1 - \phi_{(t+6)})$ is bounded by $C(1 + t)$ by Cor. 3.5.15, the second term by induction.

This shows the claim from the beginning of the proof.

An analogous argument yields that for any $T > 0$ there are $C, \varepsilon > 0$ such that the family
$$P_1 e^{-\frac{u_i}{2} t D(\rho)^2} \mathcal{R} P_1 e^{-u_{i+1} t D(\rho)^2} \mathcal{R} \ldots P_1 e^{-u_k t D(\rho)^2} \phi_t$$
is a family of Hilbert-Schmidt operators with Hilbert-Schmidt norm bounded by $Ce^{-\varepsilon t}$ for all $u_i > \frac{1}{k+1}$, $0 \leq u_{i+1}, \ldots, u_k \leq 1$ and $t > T$.

It follows that the integral
$$\int_{\Delta^k} \phi_t P_1 D(\rho)^\nu e^{-u_0 t D(\rho)^2} \mathcal{R} P_1 e^{-u_1 t D(\rho)^2} \mathcal{R} \ldots \mathcal{R} P_1 e^{-u_k t D(\rho)^2} \phi_t \, du_0 \ldots du_k$$
converges in the trace class norm and is bounded by $Ce^{-\varepsilon t}$ for some $C, \varepsilon > 0$ and all $t > T$. □

We have that
$$\phi_t D(\rho)^\nu e^{-A(\rho)_t^2} \phi_t = \sum_{k=0}^\infty (-1)^k t^{k/2} \, \phi_t D(\rho)^\nu I_k(\rho, t) \phi_t \ .$$

The decomposition $e^{-u_i t D(\rho)^2} = P_0 + P_1 e^{-u_i t D(\rho)^2}$ induces a decomposition of $D(\rho)^\nu I_k(\rho, t)$ into a sum of 2^{k+1} terms. For $0 \leq j \leq k+1$ let $P_{jk}^\nu(t)$ be the sum of those terms with exactly j factors of the form $P_1 e^{-u_i t D(\rho)^2}$.

Thus
$$D(\rho)^\nu e^{-A(\rho)_t^2} = \sum_{k=0}^\infty (-1)^k t^{\frac{k}{2}} \sum_{j=0}^{k+1} P_{jk}^\nu(t) \ .$$

Note that Prop. 3.5.5 implies that P_0 is a trace class operator, hence the operators $P_{jk}^\nu(t)$ with $j \neq k+1$ are trace class for $t \neq 0$.

From $P_0 \mathcal{R} P_0 = P_0 [W^* \, dW, D(\rho)]_s P_0 = 0$ it follows that $P_{jk}^\nu(t) = 0$ for $j < \frac{k}{2}$.
Furthermore for k even
$$P_{\frac{k}{2}k}^\nu(t) = \int_{\Delta^k} D(\rho)^\nu P_0 \mathcal{R} P_1 e^{-u_1 t D(\rho)^2} \mathcal{R} P_0 \mathcal{R} P_1 e^{-u_3 t D(\rho)^2} \mathcal{R} P_0 \ldots$$
$$\ldots P_0 \mathcal{R} P_1 e^{-u_{k-1} t D(\rho)^2} \mathcal{R} P_0 \, du_0 \ldots du_k \ .$$

Since $D(\rho) P_0 = 0$, it follows that $P_{\frac{k}{2}k}^1(t) = 0$.

Moreover by the previous lemma
$$\|\phi_t P_{(k+1)k}^\nu(t) \phi_t\|_{TR} \leq Ce^{-\varepsilon t}$$
for t large.

Now we study the behavior of the remaining cases for large t.

LEMMA 4.4.6. *Let $\nu = 0, 1$ and $j, k \in \mathbb{N}_0$ with $\frac{k}{2} \leq j \leq k$. For every $T > 0$ there is $C > 0$ such that for $t > T$*
$$\|P_{jk}^\nu(t)\|_{TR} \leq C t^{-j} \ .$$
If k is odd, then for every $n \in \mathbb{N}$ and $T > 0$ there is $C > 0$ such that for $t > T$
$$|\mathrm{Tr}_s \phi_t^2 P_{\frac{k+1}{2}k}^\nu(t)| \leq C t^{-n} \ .$$

PROOF. For $j \leq k$ the operator $P_{jk}^\nu(t)$ is a sum of terms of the form
$$\int_{\Delta^k} (A(u_0, \ldots u_i, t) P_0)(P_0 B(u_{i+1}, \ldots u_k, t)) \, du_0 \ldots du_k \ ,$$

4.4. THE INDEX THEOREM AND ITS PROOF

where A and B are continuous families of bounded operators on $L^2(M, E \otimes \hat{\Omega}_{\leq \mu} \mathcal{A}_i)$ for $u_0 \neq 0$.

Since P_0 is a Hilbert-Schmidt operator by Prop. 3.5.5, Prop. 5.2.13 implies that

$$\| \int_{\Delta^k} A(u_0, \ldots u_i, t) P_0 \big) \big(P_0 B(u_{i+1}, \ldots u_k, t) \, du_0 \ldots du_k \|_{TR}$$

$$\leq C \| P_0 \|_{HS}^2 \int_{\Delta^k} \| A(u_0, \ldots u_i, t) \| \| B(u_{i+1}, \ldots u_k, t) \| \, du_0 \ldots du_k \ .$$

Let $\omega > 0$ be such that there is $C > 0$ with $\| P_1 e^{-tD(\rho)^2} \| \leq C e^{-\omega t}$ for all $t > 0$. Then

$\| P_{jk}^\nu(t) \|_{TR}$

$$\leq C \int_0^1 du_0 \, (u_0 t)^{-\nu/2} e^{-\omega u_0 t} \int_{(1-u_0) \Delta^{k-1}} \exp(-\sum_{i=1}^{j-1} \omega u_i t) \, du_1 \ldots du_k$$

$$= C \int_0^1 du_0 \, (u_0 t)^{-\nu/2} e^{-\omega u_0 t} \int_0^{1-u_0} e^{-\omega s t} \operatorname{vol}(s \Delta^{j-2}) \operatorname{vol}\big((1 - u_0 - s) \Delta^{k-j}\big) \, ds$$

$$= C \int_0^1 du_0 \, (u_0 t)^{-\nu/2} e^{-\omega u_0 t} \int_0^{1-u_0} e^{-\omega s t} \frac{s^{j-2}(1 - u_0 - s)^{k-j}}{(j-2)!(k-j)!} ds$$

$$= C t^{-j} \int_0^t dy \, y^{-\nu/2} e^{-\omega y} \int_0^{t-y} e^{-\omega x} \frac{x^{j-2}(1 - y/t - x/t)^{k-j}}{(j-2)!(k-j)!} \, dx$$

$$\leq C t^{-j} \Big(\int_0^\infty y^{-\nu/2} e^{-\omega y} \, dy \Big) \Big(\int_0^\infty e^{-\omega x} \frac{x^{j-2}}{(j-2)!(k-j)!} \, dx \Big)$$

$$\leq C t^{-j} \ .$$

This shows the first statement.

For k odd

$\operatorname{Tr}_s \phi_t^2 P_{\frac{k+1}{2}k}^\nu(t)$

$$= \int_{\Delta^k} \operatorname{Tr}_s \phi_t^2 D(\rho)^\nu P_0 \mathcal{R} P_1 e^{-u_1 tD(\rho)^2} \mathcal{R} P_0 \ldots P_0 \mathcal{R} P_1 e^{-u_k tD(\rho)^2} \, du_0 \ldots du_k$$

$$+ \int_{\Delta^k} \operatorname{Tr}_s D(\rho)^\nu P_1 e^{-u_0 tD(\rho)^2} \mathcal{R} P_0 \mathcal{R} P_1 \ldots e^{-u_{k-1} tD(\rho)^2} \mathcal{R} P_0 \phi_t^2 \, du_0 \ldots du_k \ .$$

By Prop. 3.5.5 we can estimate $\|(1 - \phi_t^2) P_0\|_{HS}$ and $\| P_0(1 - \phi_t^2) \|_{HS}$ by $C t^{-n}$. The second estimate follows then from the cyclicity of the supertrace since $P_0 P_1 = P_1 P_0 = 0$. $\square$

From the estimates so far obtained the second assertion of the theorem follows.

Furthermore the previous lemmas imply that for $t > T$ there is $C > 0$ such that

$$|\operatorname{Tr}_s \phi_t^2 e^{-A(\rho)_t^2} - \sum_{n=0}^\infty t^n \operatorname{Tr}_s \phi_t^2 P_{n(2n)}^0(t)| \leq C t^{-\frac{1}{2}} \ .$$

Hence it remains to study the behavior of $t^n P_{n(2n)}(t)$ for $t \to \infty$.

Recall that

$$P_{n(2n)}^0(t) = \int_{\Delta^{2n}} P_0 \mathcal{R} P_1 e^{-u_0 tD(\rho)^2} \mathcal{R} P_0 \mathcal{R} P_1 e^{-u_1 tD(\rho)^2} \mathcal{R} P_0 \ldots$$

$$\ldots P_0 \mathcal{R} P_1 e^{-u_{n-1} tD(\rho)^2} \mathcal{R} P_0 \, du_0 \ldots du_{2n} \ .$$

By Prop. 3.5.5 for any $j \in \mathbb{N}$ there is $C > 0$ such that
$$\|P_0 - \phi_t^2 P_0\|_{HS} \leq Ct^{-j} ,$$
hence
$$|\text{Tr}_s \phi_t^2 P_{n(2n)}^0(t) - \text{Tr}_s P_{n(2n)}^0(t))| \leq Ct^{-j} .$$
In the next lemma we show that for $T > 0$ there is $C > 0$ such that
$$\|t^n P_{n(2n)}^0(t) - (-1)^n \frac{1}{n!}(P_0 W dW^* P_0)^{2n}\| \leq Ct^{-1}$$
in $B(L^2(M, E \otimes \hat{\Omega}_{\leq \mu} \mathcal{A}_i))$ for $t > T$. This implies that there is $C > 0$ such that for $t > T$
$$|t^n \text{Tr}_s P_{n(2n)}^0(t) - (-1)^n \frac{1}{n!} \text{Tr}_s (P_0 W dW^* P_0)^{2n}|$$
$$\leq C\|P_0\|_{HS}^2 \|t^n P_{n(2n)}^0(t) - (-1)^n \frac{1}{n!}(P_0 W dW^* P_0)^{2n}\| \leq Ct^{-1} .$$
Both estimates combined yield the first assertion of the theorem.

LEMMA 4.4.7. *Let $k, n \in \mathbb{N}_0$ with $n \leq k$. For $t \to \infty$ the term*
$$t^n \int_{\Delta^k} P_0 \mathcal{R} P_1 e^{-u_0 t D(\rho)^2} P_1 \mathcal{R} P_0 \mathcal{R} P_1 e^{-u_1 t D(\rho)^2} \mathcal{R} P_0 \ldots$$
$$\ldots P_0 \mathcal{R} P_1 e^{-u_{n-1} t D(\rho)^2} \mathcal{R} P_0 \, du_0 \ldots du_k$$
converges in $B(L^2(M, E \otimes \hat{\Omega}_{\leq \mu} \mathcal{A}_i))$ to
$$(-1)^n \frac{1}{(k-n)!} (P_0 W \, dW^* P_0)^{2n}$$
with $O(t^{-1})$.

PROOF. For $i, j \in \{0, 1\}$ write ${}_i d_j := P_i W \, dW^* P_j$.
By $\mathcal{R} = [W^* \, dW, D(\rho)]_s$
$$P_1 \mathcal{R} P_0 = P_1 D(\rho) W^* \, dW P_0 ,$$
$$P_0 \mathcal{R} P_1 = P_0 W^* \, dW D(\rho) P_1 ,$$
thus
$$P_0 \mathcal{R} P_1 e^{-t D(\rho)^2} \mathcal{R} P_0 = {}_0 d_1 D(\rho)^2 e^{-t D(\rho)^2} {}_1 d_0 .$$
This term is uniformly bounded for $t \to 0$. Hence the integral
$$\int_0^t {}_0 d_1 D(\rho)^2 e^{-s D(\rho)^2} {}_1 d_0 \, ds$$
converges and equals ${}_0 d_1 (e^{-t D(\rho)^2} - 1) \, {}_1 d_0$.
For $n \in \mathbb{N}$ and $k \geq n$
$$t^n \int_{\Delta^k} P_0 \mathcal{R} P_1 e^{-u_0 t D(\rho)^2} \mathcal{R} P_0 \mathcal{R} P_1 e^{-u_1 t D(\rho)^2} \mathcal{R} P_0 \ldots$$
$$\ldots P_0 \mathcal{R} P_1 e^{-u_{n-1} t D(\rho)^2} \mathcal{R} P_0 \, du_0 \ldots du_k$$
$$= t^n \int_{\Delta^k} {}_0 d_1 D(\rho)^2 e^{-u_0 t D(\rho)^2} {}_1 d_0 {}_0 d_1 D(\rho)^2 e^{-u_1 t D(\rho)^2} {}_1 d_0 \ldots$$
$$\ldots {}_0 d_1 D(\rho)^2 e^{-u_{n-1} t D(\rho)^2} {}_1 d_0 \, du_0 \ldots du_k .$$

4.4. THE INDEX THEOREM AND ITS PROOF

Set
$$D_n := \{(u_0, u_1, \ldots u_{n-1}) \in \mathbb{R}^n \mid \sum_{i=0}^{n-1} u_i \leq 1;\ 0 \leq u_i \leq 1,\ i = 0, \ldots, n-1\}.$$

By integration on $u_n, \ldots, u_k$ the previous term equals

$$(*) \quad t^n \int_{D_n} \ldots {}_0 d_1 D(\rho)^2 e^{-u_{n-1} t D(\rho)^2} {}_1 d_0\ \mathrm{vol}((1 - \sum_{i=0}^{n-1} u_i) \Delta^{k-n})\ du_0 \ldots du_{n-1}.$$

We claim that $(*)$ converges to $\frac{1}{(k-n)!}({}_0 d_{11} d_0)^n$ with $O(t^{-1})$. Then it follows from $(W^* \mathrm{d} W)^2 = 0$ that ${}_0 d_{11} d_0 = -{}_0 d_0^2$. This shows the assertion of the lemma.

For $n=1$ and $k=1$ the term $(*)$ equals 1.

For $n=1$ and $k>1$ the claim follows by induction since by partial integration

$$\int_0^1 {}_0 d_1 D(\rho)^2 e^{-u_0 t D(\rho)^2} {}_1 d_0\ \mathrm{vol}((1-u_0)\Delta^{k-1})\ du_0$$
$$= {}_0 d_1 e^{-tD(\rho)^2} {}_1 d_0 + \int_0^1 {}_0 d_1 D(\rho)^2 e^{-u_0 t D(\rho)^2} {}_1 d_0\ \mathrm{vol}((1-u_0)\Delta^{k-2})\ du_0$$

because of
$$\partial_{u_0} \mathrm{vol}((1-u_0)\Delta^{k-1}) = -\mathrm{vol}((1-u_0)\Delta^{k-2}).$$

Furthermore ${}_0 d_1 e^{-tD(\rho)^2} {}_1 d_0$ decays exponentially for $t \to \infty$.

For general k and n the term $(*)$ equals, by partial integration on u_{n-1},

$$t^n \int_{\Delta^{n-1}} \ldots {}_0 d_1 \left[-t^{-1} e^{-xtD(\rho)^2} \mathrm{vol}((1-\sum_{i=0}^{n-2} u_i - x)\Delta^{k-n}) \right]_0^{u_{n-1}} {}_1 d_0\ du_0 \ldots du_{n-1}$$

$$+\ t^{n-1} \int_{D_n} \ldots {}_0 d_1 e^{-u_{n-1} t D(\rho)^2} {}_1 d_0\ \partial_{u_{n-1}} \mathrm{vol}((1-\sum_{i=0}^{n-1} u_i)\Delta^{k-n})\ du_0 \ldots du_{n-1}.$$

Note that the first integral vanishes for $x = u_{n-1}$.

We obtain

$$t^{n-1} \int_{D_{n-1}} \ldots {}_0 d_1 D(\rho)^2 e^{-u_{n-2} t D(\rho)^2} {}_1 d_0 {}_0 d_{11} d_0\ \mathrm{vol}((1-\sum_{i=0}^{n-2} u_i)\Delta^{k-n})\ du_0 \ldots du_{n-2}$$

$$-t^{n-1} \int_{D_{n-1}} \ldots {}_0 d_1 e^{-u_{n-1} t D(\rho)^2} {}_1 d_0\ \mathrm{vol}((1-\sum_{i=0}^{n-1} u_i)\Delta^{k-n-1})\ du_0 \ldots du_{n-1}.$$

There are $C, \omega > 0$ such that the last term is bounded by

$$Ct^{n-1} \int_0^1 e^{-s\omega t}\ \mathrm{vol}(s\Delta^{n-1})\ \mathrm{vol}((1-s)\Delta^{k-n-1})\ ds,$$

hence it vanishes with $O(t^{-1})$ for $t \to \infty$ (by a calculation as the proof of the previous lemma).

By induction the first term converges with $O(t^{-1})$ to
$$\frac{1}{(k-1-(n-1))!}({}_0 d_{11} d_0)^n = \frac{1}{(k-n)!}({}_0 d_{11} d_0)^n.$$

$\square$

4.4.3. The limit of $\operatorname{Tr}_s e^{-A_t^2}$ for $t \to 0$. Recall the definition of $\mathcal{N}$ from §2.5.

THEOREM 4.4.8. (1) $\displaystyle\lim_{t\to 0} \operatorname{Tr}_s e^{-A_t^2} = \mathcal{N}$.

(2) $\displaystyle\lim_{t\to 0} \operatorname{Tr}_s D e^{-A_t^2} = 0$.

PROOF. (1) By Prop. 4.3.1

$$\lim_{t\to 0} \operatorname{Tr}_s e^{-A_t^2} = \sum_{n=0}^\infty (-1)^n \lim_{t\to 0} t^{\frac{n}{2}} \sum_{k\in J} \int_{\mathcal{U}_k} \gamma_k(x) \operatorname{tr}_s w(0)_t^k(x,x)^n \phi_k(x) \, dx \ .$$

Now $\operatorname{tr}_s w(\rho)_t^k(x,x)^n = 0$ for $k \in \mathbb{Z}/6$ and for all $n \in \mathbb{N}_0$ and $t > 0$ by §4.2.

Furthermore $w(0)_t^\clubsuit(x,y)^n = 0$ for $n > 0$.

Recall that $\mathcal{U}_\clubsuit$ contains the isolated point $*$. Since $w(0)_t^\clubsuit(x,y)^0$ is the integral kernel of $e^{-tD_N^2}$,

$$\lim_{t\to 0} \operatorname{tr}_s w(0)_t^\clubsuit(x,x)^0 = \mathcal{N} 1_*(x)$$

in $C(\mathcal{U}_\clubsuit)$ by the local index theorem ([**BGV**], Th. 4.2) and by $\operatorname{ch}(E/S) = \operatorname{ch}((\mathbb{C}^+)^d) + \operatorname{ch}((\mathbb{C}^-)^d) = 0$.

(2) Prop. 4.3.1 implies that

$$\lim_{t\to 0} \operatorname{Tr}_s D e^{-A_t^2} = \sum_{n=0}^\infty (-1)^n \lim_{t\to 0} t^{\frac{n}{2}} \operatorname{Tr}_s D W_n(0,t) \ .$$

Since $DW_0(0,t)$ is an odd trace class operator, its supertrace vanishes. Furthermore $\operatorname{Tr}_s DW_n(0,t) = 0$ by §4.2. $\square$

4.4.4. $\dfrac{d}{dt}\operatorname{Tr}_s e^{-A(\rho)_t^2}$ **and** $\dfrac{d}{d\rho}\operatorname{Tr}_s e^{-A(\rho)_t^2}$.

LEMMA 4.4.9. (1)

$$\frac{d}{dt}\operatorname{Tr}_s e^{-A(\rho)_t^2} = -\operatorname{Tr}_s \frac{dA(\rho)_t^2}{dt} e^{-A(\rho)_t^2} \ .$$

(2)

$$\frac{d}{d\rho}\operatorname{Tr}_s e^{-A(\rho)_t^2} = -\operatorname{d} \operatorname{Tr}_s \frac{dA(\rho)_t}{d\rho} e^{-A(\rho)_t^2} \ .$$

PROOF. (1) First we calculate $\frac{d}{dt} e^{-A(\rho)_t^2}$.

Consider the holomorphic semigroup $e^{-t'(D(\rho)^2 + z\mathcal{R})}$ depending on the parameter z. By the semigroup law

$$\frac{d}{dt'} e^{-t'(D(\rho)^2 + z\mathcal{R})} = -(D(\rho)^2 + z\mathcal{R}) e^{-t'(D(\rho)^2 + z\mathcal{R})}$$

and by Duhamel's formula (Prop. 5.4.4)

$$\frac{d}{dz} e^{-t'(D(\rho)^2 + z\mathcal{R})} = -\int_0^{t'} e^{-(t'-s)(D(\rho)^2 + z\mathcal{R})} \mathcal{R} e^{-s(D(\rho)^2 + z\mathcal{R})} ds \ .$$

4.4. THE INDEX THEOREM AND ITS PROOF

It follows that

$$\begin{aligned}
\frac{d}{dt} e^{-A(\rho)_t^2} &= \frac{d}{dt} e^{-t(D(\rho)^2 + t^{-1/2}\mathcal{R})} \\
&= \frac{d}{dt'} e^{-t'(D(\rho)^2 + t^{-1/2}\mathcal{R})}\big|_{(t'=t)} - \frac{1}{2} t^{-3/2} \frac{d}{dz} e^{-t(D(\rho)^2 + z\mathcal{R})}\big|_{(z=t^{-1/2})} \\
&= -(D(\rho)^2 + t^{-1/2}\mathcal{R}) e^{-t(D(\rho)^2 + t^{-1/2}\mathcal{R})} \\
&\quad + \frac{1}{2} t^{-3/2} \int_0^t e^{-(t-s)(D(\rho)^2 + t^{-1/2}\mathcal{R})} \mathcal{R} e^{-s(D(\rho)^2 + t^{-1/2}\mathcal{R})} ds \\
&= -t^{-1} A(\rho)_t^2 e^{-A(\rho)_t^2} + \frac{1}{2} t^{-1/2} \int_0^1 e^{-(1-s)A(\rho)_t^2} \mathcal{R} e^{-sA(\rho)_t^2} ds .
\end{aligned}$$

Using this equation we prove that

$$(*_1) \qquad |\frac{d}{dt} \operatorname{tr}_s p(\rho)_t(x,x)^n| \leq C e^{-cd(x,M_r)^2}$$

for all x uniformly in t for $t \in [t_1, t_2]$ with $t_1, t_2 > 0$. This yields that

$$\frac{d}{dt} \operatorname{Tr}_s e^{-A(\rho)_t^2} = \operatorname{Tr}_s \frac{d}{dt} e^{-A(\rho)_t^2} .$$

By Cor. 4.3.3 and the fact that the pointwise supertrace of the integral kernel $(A_t^{Z_k})^2 e^{-(A_t^{Z_k})^2}$ vanishes for $k \in \mathbb{Z}/6$ the pointwise supertrace of the integral kernel of $A(\rho)_t^2 e^{-A(\rho)_t^2}$ can be estimated by $Ce^{-cd(x,M_r)^2}$ uniformly in t on compact subsets of $(0,\infty)$.

The integral kernel of $\int_0^1 e^{-(1-s)A(\rho)_t^2} \mathcal{R} e^{-sA(\rho)_t^2} ds$ is the sum over $m,n \in \mathbb{N}_0$ of the terms

$$(-1)^{m+n} t^{\frac{m+n}{2}} \int_0^1 (1-s)^{m/2} s^{n/2} \int_M p(\rho)_{(1-s)t}(x,y)^m \mathcal{R}(y) p(\rho)_{st}(y,x)^n \, dy ds .$$

We decompose

$$\begin{aligned}
& p(\rho)_{(1-s)t}(x,y)^m \mathcal{R}(y) p(\rho)_{st}(y,x)^n \\
&= p(\rho)_{(1-s)t}(x,y)^m \mathcal{R}(y) \big(p(\rho)_{st}(y,x)^n - w(\rho)_{st}(y,x)^n \big) \\
&\quad + \big(p(\rho)_{(1-s)t}(x,y)^m - w(\rho)_{(1-s)t}(x,y)^m \big) \mathcal{R}(y) w(\rho)_{st}(y,x)^n \\
&\quad + w(\rho)_{(1-s)t}(x,y)^m \mathcal{R}(y) w(\rho)_{st}(y,x)^n .
\end{aligned}$$

By Prop. 4.3.1 and the fact that the operator $I_m(\rho, (1-s)t)$ is uniformly bounded on $C_R(M, E \otimes \hat{\Omega}_{\leq \mu} \mathcal{A}_i)$ for $0 \leq (1-s)t \leq t_2$ there are $C, c, r > 0$ such that for all $x \in M$ and $0 \leq s \leq 1$ and $t_1 \leq t \leq t_2$

$$|\int_M p(\rho)_{(1-s)t}(x,y)^m \mathcal{R}(y) \big(p(\rho)_{st}(y,x)^n - w(\rho)_{st}(y,x)^n \big) dy| \leq C e^{-cd(x,M_r)^2} .$$

An analogous estimate holds for the second term.

By §4.2 and since $\mathcal{R}|_{\mathcal{U}_*} = 0$

$$\operatorname{tr}_s w(\rho)_{(1-s)t}(x,y)^m \mathcal{R}(y) w(\rho)_{st}(y,x)^n = 0 .$$

We conclude that there are $r, c, C > 0$ such that for $x \in M_r$, $0 \leq s \leq 1$ and $t_1 \leq t \leq t_2$

$$(*_2) \qquad |\operatorname{tr}_s \int_M p(\rho)_{(1-s)t}(x,y)^m \mathcal{R}(y) p(\rho)_{st}(y,x)^n \, dy| \leq C e^{-cd(x,M_r)^2} .$$

Now $(*_1)$ follows.

The next step is to show that
$$\mathrm{Tr}_s \int_0^1 e^{-(1-s)A(\rho)_t^2} \mathcal{R} e^{-sA(\rho)_t^2} ds = \mathrm{Tr}_s \mathcal{R} e^{-A(\rho)_t^2}$$
or equivalently
$$\int_M \int_0^1 (1-s)^{m/2} s^{n/2} \int_M \mathrm{tr}_s p(\rho)_{(1-s)t}(x,y)^m \mathcal{R}(y) p(\rho)_{st}(y,x)^n \, dy ds dx$$
$$= \int_0^1 (1-s)^{m/2} s^{n/2} \int_M \int_M \mathrm{tr}_s p(\rho)_{(1-s)t}(x,y)^m \mathcal{R}(y) p(\rho)_{st}(y,x)^n \, dx dy ds \; .$$
We can interchange the integration over s and x by the estimate $(*_2)$.

Fix s and t. Consider once more the decomposition of
$$p(\rho)_{(1-s)t}(x,y)^m \mathcal{R}(y) p(\rho)_{st}(y,x)^n$$
from above. From Cor. 4.3.2 it follows that for $\varepsilon > 0$ there are $r, c, C > 0$ such that for all $x, y \in M$
$$|p(\rho)_{(1-s)t}(x,y)^m \mathcal{R}(y) \big(p(\rho)_{st}(y,x)^n - w(\rho)_{st}(y,x)^n\big)|$$
$$\leq C\big(e^{-cd(x,y)^2} + e^{-cd(x,M_r)^2}\big) e^{-cd(y,M_r)^2} \; .$$
The second term can be estimated in an analogous manner and the supertrace of the third term vanishes as we saw.

Hence we can interchange dx and dy.

This shows (1) by
$$\mathrm{Tr}_s \frac{d}{dt} e^{-A(\rho)_t^2} = \mathrm{Tr}_s (-t^{-1} A(\rho)_t^2 + \frac{1}{2} t^{-1/2} \mathcal{R}) e^{-A(\rho)_t^2}$$
$$= -\mathrm{Tr}_s \frac{dA(\rho)_t^2}{dt} e^{-A(\rho)_t^2} \; .$$

(2) As above, since $A(\rho)_t^2 = t(D^2 + \rho[D,K]_s + \rho^2 K^2) + \sqrt{t}\mathcal{R}$, Duhamel's formula (Prop. 5.4.4) and the chain rule imply that
$$\frac{d}{d\rho} e^{-A(\rho)_t^2} = -\int_0^1 e^{-(1-s)A(\rho)_t^2} \frac{dA(\rho)_t^2}{d\rho} e^{-sA(\rho)_t^2} \, ds \; .$$
Note that $\frac{dA(\rho)_t^2}{d\rho}$ is a trace class operator, hence we conclude immediately that
$$\frac{d}{d\rho} \mathrm{Tr}_s e^{-A(\rho)_t^2} = \mathrm{Tr}_s \frac{d}{d\rho} e^{-A(\rho)_t^2} = -\mathrm{Tr}_s \frac{dA(\rho)_t^2}{d\rho} e^{-A(\rho)_t^2} \; .$$
Since $A(\rho)_t$ is a bounded perturbation of $\sqrt{t} W D_s W^*$, we have, by Cor. 3.5.8 and Prop. 5.4.10, that $A(\rho)_t$ and $e^{-A(\rho)_t^2}$ commute.

It follows that
$$\frac{dA(\rho)_t^2}{d\rho} e^{-A(\rho)_t^2} = [A(\rho)_t, \frac{dA(\rho)_t}{d\rho} e^{-A(\rho)_t^2}]_s$$
$$= [W^* dW, \frac{dA(\rho)_t}{d\rho} e^{-A(\rho)_t^2}]_s + \sqrt{t}[D(\rho), \frac{dA(\rho)_t}{d\rho} e^{-A(\rho)_t^2}]_s \; .$$
The first term of the last line is an integral operator with integral kernel $W^* d(k) W$ where k is the integral kernel of $W \frac{dA(\rho)_t}{d\rho} e^{-A(\rho)_t^2} W^*$. Hence the supertrace of the

first term equals
$$\mathrm{d}\,\mathrm{Tr}_s \frac{dA(\rho)_t}{d\rho} e^{-A(\rho)_t^2}.$$

Now consider the second term. Let P be the orthogonal projection onto the range of K. It is an even Hilbert-Schmidt operator with smooth complex integral kernel. Since $\frac{dA(\rho)_t}{d\rho} = \sqrt{t}K$,

$$\mathrm{Tr}_s[D(\rho), \frac{dA(\rho)_t}{d\rho}e^{-A(\rho)_t^2}]_s = \mathrm{Tr}_s\sqrt{t}[D(\rho)P, Ke^{-A(\rho)_t^2}]_s.$$

Since $D(\rho)P$ and $Ke^{-A(\rho)_t^2}$ are Hilbert-Schmidt operators, the supertrace vanishes. $\square$

Let D_{I_k} be the operator D_I from §3.3.1 with boundary conditions given by the pair $(\mathcal{P}_{k \bmod 3}, \mathcal{P}_{k+1 \bmod 3})$.

In §4.3.2 we defined $A_t^{Z_k} = \tilde{W}_k^* \mathrm{d}\, \tilde{W}_k + \sqrt{t}D_{Z_k}$ such that $A_t^{Z_k}$ coincides with $A(\rho)_t$ on $\mathcal{U}_k$. There is $U_k \in C^\infty([0,1], M_{2d}(\mathcal{A}_\infty))$ such that $\tilde{W}_k(x_1, x_2) = U_k(x_2) \oplus U_k(x_2)$. Let $A_t^{I_k} = U_k^* \mathrm{d}\, U_k + \sqrt{t}\sigma D_{I_k}$.

LEMMA 4.4.10.
$$\frac{d}{dt}\mathrm{Tr}_s e^{-A(\rho)_t^2} = \frac{1}{\sqrt{4\pi t}}\sum_{k\in\mathbb{Z}/6}\mathrm{Tr}_\sigma D_{I_k}e^{-(A_t^{I_k})^2} - \frac{1}{2\sqrt{t}}\mathrm{d}\,\mathrm{Tr}_s D(\rho)e^{-A(\rho)_t^2}.$$

PROOF. Since $\frac{dA(\rho)_t^2}{dt} = [\frac{dA(\rho)_t}{dt}, A(\rho)_t]_s$, it follows from the previous lemma that
$$\frac{d}{dt}\mathrm{Tr}_s e^{-A(\rho)_t^2} = -\mathrm{Tr}_s[\frac{dA(\rho)_t}{dt}, A(\rho)_t]_s e^{-A(\rho)_t^2}.$$

Furthermore
$$-[\frac{dA(\rho)_t}{dt}, A(\rho)_t]_s e^{-A(\rho)_t^2}$$
$$= -[A(\rho)_t, \frac{dA(\rho)_t}{dt}e^{-A(\rho)_t^2}]_s$$
$$= -\frac{1}{2\sqrt{t}}[W^*\mathrm{d}\,W, D(\rho)e^{-A(\rho)_t^2}]_s - \frac{1}{2}[D(\rho), D(\rho)e^{-A(\rho)_t^2}]_s.$$

The supertrace of the first supercommutator in the last line equals
$$-\frac{1}{2\sqrt{t}}\mathrm{d}\,\mathrm{Tr}_s D(\rho)e^{-A(\rho)_t^2}.$$

Now consider the second term.

We have that
$$-\frac{1}{2}\mathrm{Tr}_s[D(\rho), D(\rho)e^{-A(\rho)_t^2}]_s$$
$$= -\frac{1}{2}\lim_{r\to\infty}\left(\mathrm{Tr}_s\phi_r D(\rho)^2 e^{-A(\rho)_t^2}\phi_r + \mathrm{Tr}_s\phi_r D(\rho)e^{-A(\rho)_t^2}D(\rho)\phi_r\right).$$

Since for $\nu \in \mathbb{N}_0$ the operators $\phi_r D(\rho)^\nu e^{-A(\rho)_t^2/2}$ and $e^{-A(\rho)_t^2/2} D(\rho)^\nu \phi_r$ are Hilbert-Schmidt operators, it follows that

$$-\frac{1}{2}\operatorname{Tr}_s[D(\rho), D(\rho)e^{-A(\rho)_t^2}]_s$$
$$= -\frac{1}{2}\lim_{r\to\infty}(\operatorname{Tr}_s e^{-A(\rho)_t^2/2}\phi_r^2 D(\rho)^2 e^{-A(\rho)_t^2/2} - \operatorname{Tr}_s e^{-A(\rho)_t^2/2} D(\rho)\phi_r^2 D(\rho)e^{-A(\rho)_t^2/2})$$
$$= \frac{1}{2}\lim_{r\to\infty}\operatorname{Tr}_s e^{-A(\rho)_t^2/2}c(d\phi_r^2)D(\rho)e^{-A(\rho)_t^2/2}$$
$$= \frac{1}{2}\lim_{r\to\infty}\operatorname{Tr}_s c(d\phi_r^2)D(\rho)e^{-A(\rho)_t^2} \ .$$

For $r > 0$ we define the function $\chi_r : Z \to \mathbb{R}$, $\chi_r(x) := \chi^2(x_1 - r)$ where χ is the function from the beginning of §4.4.1. Cor. 4.3.3 implies that

$$\frac{1}{2}\lim_{r\to\infty}\operatorname{Tr}_s c(d\phi_r^2)D(\rho)e^{-A(\rho)_t^2} = \frac{1}{2}\lim_{r\to\infty}\sum_{k\in\mathbb{Z}/6}\operatorname{Tr}_s c(d\chi_r)D_{Z_k}e^{-(A_t^{Z_k})^2} \ .$$

Recall from §4.2 that the integral kernel of $e^{-(A_t^{Z_k})^2}$ is $\sum_{n=0}^\infty (-1)^n t^{\frac{n}{2}} p_t^{Z_k}(x,y)^n$ with

$$p_t^{Z_k}(x,y)^n = c(dx_1)^n \frac{1}{\sqrt{4\pi t}} e^{-\frac{(x_1-y_1)^2}{4t}} \left(p_t^{I_k}(x_2,y_2)^n \oplus (-1)^n p_t^{I_k}(x_2,y_2)^n\right) \ .$$

The integral kernel of $c(d\chi_r)D_{Z_k}e^{-(A_t^{Z_k})^2}$ is

$$\sum_{n=0}^\infty (-1)^{n+1} t^{n/2} \chi'(x_1-r)(\partial_{x_1} + I\partial_{x_2}) p_t^{Z_k}(x,y)^n \ .$$

It follows that

$$\operatorname{Tr}_s c(d\chi_r)D_{Z_k}e^{-(A_t^{Z_k})^2}$$
$$= \frac{1}{\sqrt{4\pi t}}\sum_{n=0}^\infty (-1)^n t^{n/2} \int_0^1 \operatorname{tr}_s c(dx_1)^n \left((D_{I_k}p_t^{I_k})(x_2,x_2)^n \oplus\right.$$
$$\left. \oplus (-1)^{n+1}(D_{I_k}p_t^{I_k})(x_2,x_2)^n\right) dx_2 \ .$$

Comparison with Def. 4.1.6 and the subsequent remark yields that

$$\operatorname{tr}_s c(dx_1)^n \left((D_{I_k}p_t^{I_k})(x_2,x_2)^n \oplus (-1)^{n+1}(D_{I_k}p_t^{I_k})(x_2,x_2)^n\right)$$
$$= 2i^n \operatorname{tr}_\sigma \sigma^n (D_{I_k}p_t^{I_k})(x_2,x_2)^n \ .$$

It follows that

$$\operatorname{Tr}_s c(d\chi_r)D_{Z_k}e^{-(A_t^{Z_k})^2} = \frac{2}{\sqrt{4\pi t}}\sum_{n=0}^\infty (-1)^n t^n \int_0^1 \operatorname{tr}(D_{I_k}p_t^{I_k})(x_2,x_2)^{2n} dx_2$$
$$= \frac{1}{\sqrt{\pi t}}\operatorname{Tr}_\sigma D_{I_k}e^{-(A_t^{I_k})^2} \ ,$$

hence

$$-\frac{1}{2}\operatorname{Tr}_s[D(\rho), D(\rho)e^{-A(\rho)_t^2}]_s = \sum_{k\in\mathbb{Z}/6}\frac{1}{\sqrt{4\pi t}}\operatorname{Tr}_\sigma D_{I_k}e^{-(A_t^{I_k})^2} \ .$$

□

4.4.5. The index theorem.
In this section the notation is as in Prop. 2.5.1. Let A_{I_k} be as in the previous section.

THEOREM 4.4.11.

$$\mathrm{ch}(\mathrm{ind}\, D^+) = \frac{1}{2}[\sum_{k \in \mathbb{Z}/6} \eta(A_{I_k})] \in H_{ev}^{dR}(\mathcal{A}_\infty) \ .$$

Here we understand $\mathrm{ind}\, D^+$ as a class in $K_0(\mathcal{A}_\infty)$ via the isomorphism $K_0(\mathcal{A}) \cong K_0(\mathcal{A}_\infty)$ induced by the injection $\mathcal{A}_\infty \hookrightarrow \mathcal{A}$.

PROOF. Let $\rho \neq 0$.
By Prop. 2.5.1

$$\mathrm{ind}\, D^+ = [\mathrm{Ker}\, D(\rho)^2] - [\mathcal{A}^\mathcal{N}]$$

in $K_0(\mathcal{A})$. Let P_0 be the projection onto the kernel of $D(\rho)^2$.
From Prop. 3.5.5 and Prop. 5.3.6 it follows that

$$\mathrm{ind}\, D^+ = [\mathrm{Ran}_\infty P_0] - [\mathcal{A}_\infty^\mathcal{N}] = [\mathrm{Ran}_\infty WP_0W^*] - [\mathcal{A}_\infty^\mathcal{N}]$$

in $K_0(\mathcal{A}_\infty)$, hence in $H_*^{dR}(\mathcal{A}_\infty)$

$$\mathrm{ch}(\mathrm{ind}\, D^+) = \mathrm{ch}[\mathrm{Ran}_\infty WP_0W^*] - \mathcal{N}.$$

By Prop. 5.3.6 in $H_*^{dR}(\mathcal{A}_\infty)$

$$\begin{aligned}\mathrm{ch}[\mathrm{Ran}_\infty WP_0W^*] &= \sum_{n=0}^\infty (-1)^n \frac{1}{n!} \mathrm{Tr}_s(WP_0W^* \, \mathrm{d}\, WP_0W^*)^{2n} \\ &= \sum_{n=0}^\infty (-1)^n \frac{1}{n!} \mathrm{Tr}_s(P_0W^* \, \mathrm{d}\, WP_0)^{2n} \ .\end{aligned}$$

Let $T > 0$. By Th. 4.4.3 and Th. 4.4.8 in $\hat{\Omega}_*\mathcal{A}_\infty / \overline{[\hat{\Omega}_*\mathcal{A}_\infty, \hat{\Omega}_*\mathcal{A}_\infty]_s}$

$$\begin{aligned}\sum_{n=0}^\infty (-1)^n \frac{1}{n!} \mathrm{Tr}_s(P_0W^* \, \mathrm{d}\, WP_0)^{2n} - \mathcal{N} &= \lim_{t \to \infty} \mathrm{Tr}_s e^{-A(\rho)_t^2} - \lim_{t \to 0} \mathrm{Tr}_s e^{-A_t^2} \\ &= \int_T^\infty \frac{d}{dt} \mathrm{Tr}_s e^{-A(\rho)_t^2} dt \\ &\quad + \int_0^\rho \frac{d}{d\rho'} \mathrm{Tr}_s e^{-A(\rho')_T^2} d\rho' \\ &\quad + \int_0^T \frac{d}{dt} \mathrm{Tr}_s e^{-A_t^2} dt \ .\end{aligned}$$

By the results of §4.4.4, the estimate of Th. 4.4.3 and Th. 4.4.8 the integrals converge and we have:

$$\sum_{n=0}^{\infty}(-1)^n\frac{1}{n!}\mathrm{Tr}_s(P_0W^*\,dWP_0)^{2n} - \mathcal{N} = \sum_{k\in\mathbb{Z}/6}\int_0^\infty \frac{1}{\sqrt{4\pi t}}\mathrm{Tr}_s D_{I_k}e^{-(A_t^{I_k})^2}dt$$

$$-d\int_T^\infty \frac{1}{2\sqrt{t}}\mathrm{Tr}_s D(\rho)e^{-A(\rho)_t^2}dt$$

$$-d\int_0^\rho \mathrm{Tr}_s\frac{dA(\rho')_T}{d\rho'}e^{-A(\rho')_T^2}d\rho'$$

$$-d\int_0^T \frac{1}{2\sqrt{t}}\mathrm{Tr}_s D(0)e^{-A_t^2}dt \ .$$

The assertion follows. □

COROLLARY 4.4.12. *In $H_*^{dR}(\mathcal{A}_\infty)$*

$$\mathrm{ch}\,\tau(\mathcal{P}_0,\mathcal{P}_1,\mathcal{P}_2) = [\eta(\mathcal{P}_0,\mathcal{P}_1) + \eta(\mathcal{P}_1,\mathcal{P}_2) + \eta(\mathcal{P}_2,\mathcal{P}_0)] \ .$$

PROOF. For $\rho > 0$ we define a family $A(\rho,b)$ of superconnections associated to $D(\rho)$. For $0 < b < b_0$ let $W(b) \in C^\infty(M,\mathrm{End}^+ E \otimes \mathcal{A}_\infty)$ be as in §4.3.1 such that $W(b)$ is parallel on $\{x \in M \mid d(x,\partial M) \geq b\}$. As in §4.3.1 we set

$$A(\rho,b) := W(b)^*\,dW(b) + D(\rho) \ .$$

For every $k \in \mathbb{Z}/6$ this induces a family of superconnections $A_{I_k}(b)$. By Prop. 4.1.7 and the subsequent definition

$$\lim_{b\to 0}\eta(A_{I_k}(b)) = \eta(\mathcal{P}_{k\,\mathrm{mod}\,3},\mathcal{P}_{k+1\,\mathrm{mod}\,3}) \ .$$

The assertion follows now by the previous theorem and Prop. 2.4.4. □

4.5. A gluing formula for η-forms on S^1

In the following we sketch a reinterpretation of Cor. 4.4.12 as a gluing formula for η-forms on S^1 (which we identify with $\mathbb{R}/\mathbb{Z}$ as a Riemannian manifold).

Given $u \in U(\mathcal{A}_\infty^n)$ we define a projective system of vector bundles on S^1:

$$\mathcal{L}_i(u) = ([0,1] \times \mathcal{A}_i^n)/\sim$$

with $(0,v) \sim (1,uv)$. The standard $\mathcal{A}$-valued scalar product on $\mathcal{A}^n$ induces a hermitian $\mathcal{A}$-valued metric on $\mathcal{L}(u) := \mathcal{L}_0(u)$. We identify a smooth section of $\mathcal{L}_i(u)$ with a smooth function $f : \mathbb{R} \to \mathcal{A}_i^n$ satisfying $f(x+1) = uf(x)$. The trivial connection $f \mapsto f'dx$ on $\mathbb{R} \times \mathcal{A}^n$ induces a connection on $\mathcal{L}(u)$ preserving the metric. The associated Dirac operator is denoted by $\partial\!\!\!/_{\mathcal{L}(u)}$ (as are its closures on the spaces $L^2(S^1,\mathcal{L}_i(u))$ in the following).

Assume now $1 \notin \sigma(u)$. Then there is a path $w : [0,1] \to U(\mathcal{A}_\infty^n)$ with $w(0) = 1$, $w(1) = u$, and the map $f \mapsto w^*f$ defines a trivialization of $\mathcal{L}_i(u)$. Hence $\partial\!\!\!/_{\mathcal{L}(u)}$ can be identified with the operator $i\partial + w^*w'$ on the trivial bundle. One can easily deduce that $-\partial\!\!\!/_{\mathcal{L}(u)}^2$ generates a holomorphic semigroup on $L^2(S^1,\mathcal{L}_i(u))$ with integral kernel in $C^\infty(S^1 \times S^1, M_n(\mathcal{A}_\infty))$. Since $\partial\!\!\!/_{\mathcal{L}(u)}^2$ is invertible on $L^2(S^1,\mathcal{L}_i(u))$, the integral kernel vanishes exponentially for $t \to \infty$. Locally there is a trivialization of $\mathcal{L}(u)$ with respect to which $\partial\!\!\!/_{\mathcal{L}(u)}$ equals $i\partial$. By Duhamel's principle it follows that $\mathrm{Tr}\,\partial\!\!\!/_{\mathcal{L}(u)}e^{-t(\partial\!\!\!/_{\mathcal{L}(u)})^2}$ converges to zero in $\mathcal{A}_\infty/\overline{[\mathcal{A}_\infty,\mathcal{A}_\infty]}$ for $t \to 0$.

4.5. A GLUING FORMULA FOR η-FORMS ON S^1

We define a superconnection $A := w\,\mathrm{d}\,w^* + \partial\!\!\!/_{\mathcal{L}(u)}$ and the associated rescaled superconnection $A_t := w\,\mathrm{d}\,w^* + \sqrt{t}\partial\!\!\!/_{\mathcal{L}(u)}$. Then $e^{-A_t^2}$ is a well-defined integral operator with smooth integral kernel. Furthermore $\mathrm{Tr}\,\partial\!\!\!/_{\mathcal{L}(u)}e^{-A_t^2}$ has a limit for $t \to 0$, and the η-form

$$\eta(A_w) := \frac{1}{\sqrt{\pi}} \int_0^\infty t^{-1/2}\,\mathrm{Tr}\,\partial\!\!\!/_{\mathcal{L}(u)}e^{-A_t^2}\,dt \in \hat{\Omega}_*\mathcal{A}_\infty / \overline{[\hat{\Omega}_*\mathcal{A}_\infty, \hat{\Omega}_*\mathcal{A}_\infty]_s}$$

is well-defined. As in Def. 4.1.8 we can eliminate the dependence of the trivialization w by defining

$$\eta(\partial\!\!\!/_{\mathcal{L}(u)}) = \lim_{\varepsilon \to 0} \eta(A_{w_\varepsilon})$$

where w_ε is a family of trivializations with $w'_\varepsilon \subset [0, \varepsilon] \cup [1 - \varepsilon, 1]$. The existence of the limit follows from the proof of the next proposition.

In the following let

$$P(u) := \frac{1}{2}\begin{pmatrix} 1 & u^* \\ u & 1 \end{pmatrix}.$$

PROPOSITION 4.5.1. *Let u_0, u_1 be unitaries such that $u_0 - u_1$ is invertible. Then up to exact forms*

$$\eta(\partial\!\!\!/_{\mathcal{L}(u_0^*u_1)}) = \eta(P(u_0), P(u_1)) \ .$$

PROOF. For a unitary $U \in M_n(\mathcal{A}_\infty)$

$$\eta(U^*P(u_0)U, U^*P(u_1)U) = \eta(P(u_0), P(u_1)) \ .$$

Since for $U := \begin{pmatrix} 1 & 0 \\ 0 & u_0 \end{pmatrix}$ we have that $U^*P(u_0)U = P(1)$ and $U^*P(u_1)U = P(u_0^*u_1)$, we may assume $u_0 = 1$. Let $u := u_1$.

Let $V : \mathbb{R} \to U(\mathcal{A}_\infty^n)$ be a smooth path with $V(x) = u$ for $x \leq 0$ and $V(x) = -1$ for $x \geq 1$ and with $1 \notin \sigma(V)$; we define the bundle

$$L_i(V) := (\mathbb{R} \times [0,1]) \times \mathcal{A}_i^n / \sim$$

where $(x_1, 0, v) \sim (x_1, 1, V(x_1)v)$. The smooth sections of $L_i(V)$ can be identified with smooth functions $f : \mathbb{R} \times \mathbb{R} \to \mathcal{A}^n$ fulfilling $f(x_1, x_2 + 1) = V(x_1)f(x_1, x_2)$. Let $W : \mathbb{R} \times [0,1] \to U(\mathcal{A}_\infty^n)$ be smooth with $W(x_1, 0) = 1$, $W(x_1, 1) = V(x_1)$ and such that $W(x_1, x_2)$ is independent of x_2 in a small neighborhood of 0 resp. 1 and independent of x_1 for $x_1 < 0$ resp. $x_1 > 1$. Then $f \mapsto W^*f$ is a bundle isomorphism between $L_i(V)$ and the trivial bundle $(\mathbb{R} \times S^1) \times \mathcal{A}_i^n$. Since $dx_1\partial_1 + W^*dx_2\partial_2 W = dx_1\partial_1 + dx_2\partial_2 + dx_2W^*(\partial_2 W)$ is a connection preserving the metric on the trivial bundle, the operator $Wdx_1\partial_1W^* + dx_2\partial_2$ is a connection preserving the metric on $L_i(V)$. It agrees with the trivial connection for $|x_1|$ large.

Let $\partial\!\!\!/_{L_i(V)}$ be the Dirac operator associated to the connection $Wdx_1\partial_1W^* + dx_2\partial_2$ on the bundle $L_i(V)$. The index of $\partial\!\!\!/_{L_i(V)}^+$ vanishes, which can be seen as follows. For $t \in [0,1]$ let $W_t : \mathbb{R} \times [0,1] \to U(\mathcal{A}_\infty^n)$ be a homotopy with $W_0 = 1$ and $W_1 = W$ and independent of x_1 on the complement of a compact set. Then $dx_1\partial_1 + W_t^*dx_2\partial_2 W_t$ is a homotopy of connections on the trivial bundle interpolating between the trivial connection and the connection $dx_1\partial_1 + W^*dx_2\partial_2 W$. The index of the Dirac operator associated to the trivial connection vanishes, hence the index of the Dirac operator associated to $dx_1\partial_1 + W^*dx_2\partial_2 W$ vanishes as well.

The local term in the index theorem is determined by the superconnection

$$B_1 := \mathrm{d} + W\partial\!\!\!/_{L_i(V)}W^* = \mathrm{d} + c(dx_1)\partial_1 + c(dx_2)\partial_2 + c(dx_2)W(\partial_2 W^*)$$

on the trivial bundle $S \otimes \mathcal{A}^n$ on $\mathbb{R} \times S^1$. The contribution from the cylindric ends is given by $\frac{1}{2}(\eta(\partial\!\!\!/_{\mathcal{L}(-1)}) - \eta(A))$.

Compare this with the index theorem for the Dirac operator $\partial\!\!\!/_Z$ on the sections of $(\mathbb{R} \times [0,1]) \times ((\mathcal{A}^+)^{2n} \oplus (\mathcal{A}^-)^{2n})$ with boundary conditions defined by $P(1) \oplus P(1)$ and $P(V(x_1)) \oplus P(V(x_1))$. Let
$$\tilde{W} := \mathrm{diag}(1, W, 1, W) \in M_{4n}(\mathcal{A}_\infty) \ .$$
Then
$$\tilde{W}^* \partial\!\!\!/_Z \tilde{W} = \partial\!\!\!/_Z + \tilde{W}^* c(dx_1)(\partial_1 \tilde{W}) + \tilde{W}^* c(dx_2)(\partial_2 \tilde{W}) \ ,$$
and the boundary conditions transform to $P(1) \oplus P(1)$ and $P(-1) \oplus P(-1)$.

An argument analogous to the one above shows that the index of $\tilde{W}^* \partial\!\!\!/_Z^+ \tilde{W}$ vanishes. Since $\tilde{W}^* c(dx_1)(\partial_1 \tilde{W})$ is compactly supported, the index of $\tilde{W}^* \partial\!\!\!/_Z \tilde{W}$ equals the index of $\partial\!\!\!/_Z + \tilde{W}^* c(dx_2)(\partial_2 \tilde{W})$, which is the Dirac operator associated to the connection $dx_1 \partial_1 + dx_2 \partial_2 + dx_2 \tilde{W}^*(\partial_2 \tilde{W})$. Hence we can use the superconnection $B_2 := \mathrm{d} + c(dx_1)\partial_1 + c(dx_2)\partial_2 + c(dx_2)\tilde{W}^*(\partial_2 \tilde{W})$ for the index theorem, and it follows that the local term here equals the local term in the index theorem for $\partial\!\!\!/_{L(V)}$. Since the proposition is true for $u = -1$ by [**CLM**], Prop. 6.3, the assertion follows. $\square$

COROLLARY 4.5.2. *If $u_0, u_1, u_2 \in M_n(\mathcal{A}_\infty)$ are unitaries such that $u_i - u_j$ is invertible for $i \neq j$, then*
$$\mathrm{ch}\,\tau(P(u_0), P(u_1), P(u_2)) = [\eta(\partial\!\!\!/_{\mathcal{L}(u_0^* u_1)}) + \eta(\partial\!\!\!/_{\mathcal{L}(u_1^* u_2)}) + \eta(\partial\!\!\!/_{\mathcal{L}(u_2^* u_0)})] \in H_*^{dR}(\mathcal{A}_\infty) \ .$$

Assume that $P_i - P_j$ and $P_i - (1 - P_j)$ are invertible for $i \neq j$ and define
$$\begin{aligned} \tau_I(P_0, P_1, P_2) &:= \tau(P_0, P_1, P_2) + \tau(P_0, 1 - P_1, P_1) \\ &\quad + \tau(P_1, 1 - P_2, P_2) + \tau(P_2, 1 - P_0, P_0) \in K_0(\mathcal{A}) \ . \end{aligned}$$

Then from the corollary, the previous formula and the fact that $1 - P(u) = P(-u)$ it follows after some straightforward calculations:

COROLLARY 4.5.3. *In $H_*^{dR}(\mathcal{A}_\infty)$*
$$\mathrm{ch}\,\tau_I(P(u_0), P(u_1), P(u_2)) = [\eta(\partial\!\!\!/_{\mathcal{L}(-u_0^* u_1)}) + \eta(\partial\!\!\!/_{\mathcal{L}(-u_1^* u_2)}) + \eta(\partial\!\!\!/_{\mathcal{L}(-u_2^* u_0)})] \ .$$

In the following we explain how this formula is related to a gluing formula for η-invariants ([**Bu**], §1.7 and Cor. 1.20).

Consider the Clifford bundle $\mathcal{A}^n \times [0, \pi]$ on the manifold with boundary $[0, \pi]$ with Clifford multiplication i and Dirac operator $i\partial$. Its restriction to the boundary is $\mathcal{A}^n \times (\{0\} \cup \{\pi\})$ and the Clifford multiplication by the inward pointing normal vector is given by the bundle morphism $I_0 = i \cup (-i)$. Via the standard $\mathcal{A}$-valued scalar product it induces a skew-hermitian form on the bundle $\mathcal{A}^n \times (\{0\}\cup\{\pi\})$. We identify the sections of this bundle with $\mathcal{A}^{2n}$. The image of the kernel of the Dirac operator with respect to restriction to the boundary is the Lagrangian submodule $L = \{(x, x) \mid x \in \mathcal{A}^n\}$.

Now let
$$\phi_j := (1 \cup (u_j)) : \mathcal{A}^n \times (\{0\} \cup \{\pi\}) \to \mathcal{A}^n \times (* \cup *) \ .$$
Then $\phi_j(L)$ is the range of the projection $P(u_j)$.

The bundle $\mathcal{L}(-u_j^* u_{j+1})$ on S^1 can be obtained by gluing
$$(\mathcal{A}^n \times [0, \pi]) \cup_{\phi_j^{-1} \circ I_0 \circ \phi_{j+1}} (\mathcal{A}^n \times [0, \pi]) \ ;$$

and by gluing the operator $i\partial$ on one copy of $[0, \pi]$ with $-i\partial$ on the second copy one obtains $\eth_{\mathcal{L}(-u_0^* u_1)}$.

From the proposition it follows that up to exact forms

$$\eta(\eth_{\mathcal{L}(-u_j^* u_{j+1})}) = \eta(P(u_j), 1 - P(u_{j+1})) \ .$$

Since the projection $\phi_i(L)$ in $\mathcal{A}^{2n}$ is given by the projection $P(u_j)$, we have in $H_*^{dR}(\mathcal{A}_\infty)$:

$$\operatorname{ch} \tau_I(\phi_0(L), \phi_1(L), \phi_2(L)) = [\eta(\eth_{\mathcal{L}(-u_0^* u_1)}) + \eta(\eth_{\mathcal{L}(-u_1^* u_2)}) + \eta(\eth_{\mathcal{L}(-u_2^* u_0)})] \ .$$

For $\mathcal{A} = \mathbb{C}$ this is a particular case of the well-known general gluing formula.

CHAPTER 5

Definitions and Techniques

5.1. Hilbert C^*-modules

5.1.1. Bounded operators. Let $\mathcal{A}$ be a unital C^*-algebra with norm $|\cdot|$.
In order to fix notation we recall some facts about Hilbert $\mathcal{A}$-modules. References are [**La**] and [**WO**].

DEFINITION 5.1.1. *A* PRE-HILBERT $\mathcal{A}$-MODULE *is a right $\mathcal{A}$-module H with an $\mathcal{A}$-valued scalar product $\langle \, , \, \rangle : H \times H \to \mathcal{A}$; i.e.*

(1) $\langle \, , \, \rangle$ *is $\mathcal{A}$-linear in the second variable,*
(2) $\langle x, y \rangle = \langle y, x \rangle^*$ *for all $x, y \in H$,*
(3) $\langle x, x \rangle \geq 0$ *for all $x \in H$,*
(4) *if $\langle x, x \rangle = 0$, then $x = 0$.*

If H is complete with respect to the norm $\|v\| := |\langle v, v \rangle|^{1/2}$, then H is called a HILBERT $\mathcal{A}$-MODULE.

The completion of a pre-Hilbert $\mathcal{A}$-module is a Hilbert $\mathcal{A}$-module.
The right $\mathcal{A}$-module $\mathcal{A}^n$ is a Hilbert $\mathcal{A}$-module when endowed with the standard $\mathcal{A}$-valued scalar product

$$\langle a, b \rangle := \sum_{i=1}^{n} a_i^* b_i \ .$$

The right $\mathcal{A}$-module $\{(a_n)_{n \in \mathbb{N}} \subset \mathcal{A} \mid \sum_{n=1}^{\infty} a_n^* a_n \text{ converges}\}$ endowed with the $\mathcal{A}$-valued scalar product

$$\langle (a_n)_{n \in \mathbb{N}}, (b_n)_{n \in \mathbb{N}} \rangle := \sum_{n=1}^{\infty} a_n^* b_n$$

is a Hilbert $\mathcal{A}$-module and is denoted by $l^2(\mathcal{A})$. Sometimes we use $\mathbb{Z}$ as index set.

Let M be a measure space and let $\langle \, , \, \rangle$ be the standard $\mathcal{A}$-valued scalar product on $\mathcal{A}^n$. Then the Hilbert $\mathcal{A}$-module $L^2(M, \mathcal{A}^n)$ is defined in the following way: By

$$\langle f, g \rangle_{L^2} = \int_M \langle f(x), g(x) \rangle dx$$

an $\mathcal{A}$-valued scalar product is defined on the quotient of the space of simple functions on M with values in $\mathcal{A}^n$ by the space of simple functions vanishing almost everywhere. Hence the quotient is a pre-Hilbert $\mathcal{A}$-module. Its completion is the Hilbert $\mathcal{A}$-module $L^2(M, \mathcal{A}^n)$.

Let H be a Hilbert $\mathcal{A}$-module.
A submodule $U \subset H$ is called complemented if

$$U^\perp = \{x \in H \mid \langle x, u \rangle = 0 \ \forall u \in U\}$$

satisfies $U \oplus U^\perp = H$. Any projective submodule in H is complemented.

Let H_1, H_2 be Hilbert $\mathcal{A}$-modules. The elements of
$$B(H_1, H_2) = \{T : H_1 \to H_2 \mid T \text{ is continuous and } \exists T^* : H_2 \to H_1 \text{ with}$$
$$\langle Tv, w \rangle_{H_2} = \langle v, T^*w \rangle_{H_1} \ \forall v \in H_1, \ w \in H_2\}$$
are called bounded operators from H_1 to H_2. They form a Banach space with respect to the operator norm. With the composition as a product $B(H_1) := B(H_1, H_1)$ is a C^*-algebra. Note that the existence of an adjoint must be required.

A continuous $\mathcal{A}$-module map $K : H_1 \to H_2$ is called compact if it can be approximated by a linear combination of operators of the form $x \mapsto z\langle y, x\rangle$ with $y \in H_1, z \in H_2$, in the operator norm topology.

Every compact operator is adjointable.

A projection in $B(H)$ is compact if and only if it is a projection onto a projective submodule of H.

If the range of $T \in B(H_1, H_2)$ is complemented, we call its complement the cokernel $\operatorname{Coker} T$. Clearly a necessary condition for its existence is that the range of T is closed. The following proposition shows that it is sufficient:

PROPOSITION 5.1.2. *Suppose that $T \in B(H_1, H_2)$ has closed range. Then*
   (1) $\operatorname{Ker} T$ *is a complemented submodule of H_1*,
   (2) $\operatorname{Ran} T$ *is a complemented submodule of H_2*,
   (3) $T^* : H_2 \to H_1$ *also has closed range.*

PROOF. [**La**], Th. 3.2. □

5.1.2. Fredholm operators. Let H_1, H_2 be Hilbert $\mathcal{A}$-modules isomorphic to $l^2(\mathcal{A})$.

DEFINITION 5.1.3. *An operator $F \in B(H_1, H_2)$ is* FREDHOLM *if there are decompositions $H_1 = M_1 \oplus N_1$ and $H_2 = M_2 \oplus N_2$ with the following properties:*
   (1) N_1, N_2 *are projective $\mathcal{A}$-modules and M_1, M_2 are closed $\mathcal{A}$-modules.*
   (2) *The operator F is diagonal: $F = F_M \oplus F_N$ with $F_M : M_1 \to M_2$ and $F_N : N_1 \to N_2$.*
   (3) *The component $F_M : M_1 \to M_2$ is an isomorphism.*
   (4) *The projection onto N_i along M_i is adjointable for $i = 1, 2$.*

The INDEX OF F *is defined as*
$$\operatorname{ind} F := [N_1] - [N_2] \in K_0(\mathcal{A}) .$$

PROPOSITION 5.1.4. *A selfadjoint operator $F \in B(H_1, H_2)$ is Fredholm if and only if there exists an $\mathcal{A}$-linear continuous, not necessarily adjointable, map $G : H_2 \to H_1$ such that $FG - 1$ and $GF - 1$ are compact.*

PROOF. Analogous to [**MF**], Theorem 2.4. □

From the proposition it follows that if F is Fredholm, then for any compact operator K the operator $F + K$ is Fredholm.

PROPOSITION 5.1.5. *If $F : H_1 \to H_2$ is a Fredholm operator and $K : H_1 \to H_2$ is a compact operator, then*
$$\operatorname{ind} F = \operatorname{ind}(F + K) .$$

PROOF. [**MF**], Lemma 2.3. □

Another important property of Fredholm operators is the following:

PROPOSITION 5.1.6. *If $F \in B(H_1, H_2)$ is Fredholm and $\operatorname{Ran} F$ is closed, then $\operatorname{Ker} F$ and $\operatorname{Coker} F$ are projective modules.*
Hence
$$\operatorname{ind} F = [\operatorname{Ker} F] - [\operatorname{Coker} F] .$$

PROOF. Let $P_{\operatorname{Ker} F} \in B(H_1)$ resp. $P_{\operatorname{Coker} F} \in B(H_2)$ be the orthogonal projection onto the kernel resp. cokernel of F. They exist by Prop. 5.1.2. We have to prove that they are compact.

There is $E \in B(H_2, H_1)$ such that $EF = 1 - P_{\operatorname{Ker} F}$ and $FE = 1 - P_{\operatorname{Coker} F}$. Let G be a parametrix of F. It follows that
$$P_{\operatorname{Ker} F} = 1 - EF = (1 - GF)(1 - EF)$$
and
$$P_{\operatorname{Coker} F} = 1 - FE = (1 - FE)(1 - FG)$$
are compact operators. □

PROPOSITION 5.1.7. *If $F : [0,1] \to B(H_1, H_2)$ is a continuous path of Fredholm operators, then the map $[0,1] \to K_0(\mathcal{A})$, $t \mapsto \operatorname{ind} F(t)$ is constant.*

PROOF. see [**WO**], Prop. 17.3.4. □

5.1.3. Regular operators. In this section some basic facts about unbounded operators on Hilbert $\mathcal{A}$-modules are collected. Most of them and more can be found in [**La**].

Let H be a Hilbert $\mathcal{A}$-module with $\mathcal{A}$-valued scalar product $\langle \, , \, \rangle$. Let $D : \operatorname{dom} D \to H$ be a densely defined operator on H.

LEMMA 5.1.8. *If the adjoint D^* of D is densely defined, then D is closable.*

PROOF. Let $(f_n)_{n \in \mathbb{N}}$ be a sequence in $\operatorname{dom} D$ such that (f_n, Df_n) converges to $(0, f)$ in $H \oplus H$ for $n \to \infty$. Then for every $g \in \operatorname{dom} D^*$
$$\langle f, g \rangle = \lim_{n \to \infty} \langle Df_n, g \rangle = \lim_{n \to \infty} \langle f_n, D^*g \rangle = 0 .$$
Since $\operatorname{dom} D^*$ is dense in H, it follows that $f = 0$. □

If D is closed, then
$$\langle f, g \rangle_D := \langle f, g \rangle + \langle Df, Dg \rangle$$
is an $\mathcal{A}$-valued scalar product on $\operatorname{dom} D$, with respect to which $\operatorname{dom} D$ is a Hilbert $\mathcal{A}$-module, denoted by $H(D)$.

LEMMA 5.1.9. *Assume that D is closed.*
  (1) *Suppose that D has a densely defined adjoint D^*. Then $\operatorname{Ker} D^* = (\operatorname{Ran} D)^\perp$.*
  (2) *$\operatorname{Ker} D$ is complemented in $H(D)$ if and only if $\operatorname{Ker} D$ is complemented in H.*

PROOF. (1) Since for $f \in \operatorname{Ker} D^*$ and $h \in \operatorname{dom} D$
$$\langle f, Dh \rangle = \langle D^*f, h \rangle = 0 ,$$
the $\mathcal{A}$-module $\operatorname{Ker} D^*$ is a submodule of $(\operatorname{Ran} D)^\perp$.

For $g \in (\operatorname{Ran} D)^\perp$ the $\mathcal{A}$-linear functional
$$\operatorname{dom} D \to \mathcal{A}, \ f \mapsto \langle g, Df \rangle$$

vanishes. Thus $g \in \operatorname{dom} D^*$ and $D^*g = 0$.

(2) For $g \in \operatorname{Ker} D$ and $f \in \operatorname{dom} D$ the conditions $\langle f, g \rangle_D = 0$ and $\langle f, g \rangle = 0$ are equivalent. Hence if $\operatorname{Ker} D$ is complemented in $H(D)$, then $\operatorname{Ker} D$ is complemented in H.

Conversely, if $\operatorname{Ker} D$ is complemented in H, we can decompose $g \in H(D)$ in a sum $g = g_1 + g_2$ with $g_1 \in \operatorname{Ker} D$ and $g_2 \in (\operatorname{Ker} D)^\perp$. From $\operatorname{Ker} D \subset H(D)$ it follows that $g_2 = g - g_1 \in H(D)$, hence $g_2 \in (\operatorname{Ker} D)^{\perp_{H(D)}}$. □

Recall that D is called regular if it is closed with densely defined adjoint D^* and if $1 + D^*D$ has dense range, or equivalently if it is closed with densely defined adjoint and if its graph is complemented in $H \times H$.

If D is regular, then $1 + D^*D$ has a bounded inverse.

In the following the adjoint of an operator $A \in B(H(D), H)$ is denoted by $A^T \in B(H, H(D))$ in order to distinguish it from the adjoint A^* of A as an unbounded operator on H.

LEMMA 5.1.10. *Assume that D is closed.*

(1) *The operator D is regular if and only if the inclusion $\iota : H(D) \to H$ is in $B(H(D), H)$ and $(1 + D^*D)$ is selfadjoint. Then $\iota^T = (1 + D^*D)^{-1} \in B(H, H(D))$ and $(1 + D^*D)^{-\frac{1}{2}} : H \to H(D)$ is an isometry.*
(2) *Assume that D is regular and selfadjoint. Then $D \in B(H(D), H)$ and $D^T = D(1 + D^2)^{-1}$.*

PROOF. If D is regular, then for $v \in H(D)$ and $w \in H$

$$\begin{aligned}
\langle \iota v, w \rangle &= \langle v, w \rangle \\
&= \langle v, (1 + D^*D)(1 + D^*D)^{-1} w \rangle \\
&= \langle v, (1 + D^*D)^{-1} w \rangle + \langle Dv, D(1 + D^*D)^{-1} w \rangle \\
&= \langle v, (1 + D^*D)^{-1} w \rangle_D \ .
\end{aligned}$$

This shows $\iota^T = (1 + D^*D)^{-1}$.

Now the converse direction:

Let $v \in H$. Then for any $w \in \operatorname{dom}(1 + D^*D)$

$$\langle v, w \rangle = \langle \iota^T v, w \rangle_D = \langle \iota^T v, (1 + D^*D) w \rangle \ .$$

Since $(1 + D^*D)$ is selfadjoint, it follows that $\iota^T v \in \operatorname{dom}(1 + D^*D)$ and $(1 + D^*D) \iota^T v = v$.

This shows that $(1 + D^*D)$ is surjective and that ι^T is a right inverse of $(1 + D^*D)$. Since $(1 + D^*D)$ is bounded below, it is injective as well. It follows that $(1 + D^*D)$ is invertible and ι^T is its inverse.

The remaining parts are immediate.

□

PROPOSITION 5.1.11. *Let D_0 be a regular selfadjoint operator and assume $D = D_0 + V$ with $V \in B(H)$.*

(1) *Then D is closed.*
(2) *The identity map induces a continuous isomorphism from $H(D_0)$ to $H(D)$.*
(3) *$D \in B(H(D_0), H)$.*
(4) *Suppose that V is selfadjoint. Then D is regular.*

PROOF. (1) From $\operatorname{dom} D^{**} = \operatorname{dom} D_0^{**} = \operatorname{dom} D_0 = \operatorname{dom} D$ it follows that $D = D^{**}$. Thus D is closed.

Assertion (2) follows from the fact that there is $C > 0$ such that for all $f \in H(D_0)$
$$\begin{aligned}
\|f\|_D^2 &\leq \|f\|^2 + \|D_0 f\|^2 + \|\langle Vf, D_0 f\rangle\| + \|\langle D_0 f, Vf\rangle\| + \|Vf\|^2 \\
&\leq C(\|f\|^2 + \|D_0 f\|^2) + 2\|Vf\|\|D_0 f\| \\
&\leq C\|f\|_{D_0}^2 \ .
\end{aligned}$$
We applied Cauchy-Schwarz inequality.

(3) By (2) the operator $D : H(D_0) \to H$ is continuous. By the previous lemma the adjoint of $D = D_0 + V\iota : H(D_0) \to H$ is
$$D^T = D_0(1 + D_0^2)^{-1} + (1 + D_0^2)^{-1} V^* : H \to H(D_0) \ .$$

(4) By (3) the operator $D + i : H(D_0) \to H$ is an adjointable bounded operator. By [**La**], Lemma 9.7, the range of $D + i$ is closed. Thus it is complemented by Prop. 5.1.2. From Lemma 5.1.9 it follows that the cokernel of $D + i$ agrees with the kernel of $D - i$. By [**La**], Lemma 9.7, the operator $D - i$ is injective. It follows that $\operatorname{Coker}(D + i) = \{0\}$. By [**La**], Lemma 9.8, this shows that D is regular. □

It can be shown that the condition of selfadjointness for D_0 and V in statement (4) of the previous proposition can be dropped.

PROPOSITION 5.1.12. *Let D be a regular and selfadjoint operator on H with closed range.*

(1) *The cokernel of D exists and $\operatorname{Ker} D = \operatorname{Coker} D$. In particular $\operatorname{Ker} D$ is complemented.*
(2) *The $\mathcal{A}$-module $\operatorname{dom} D \cap \operatorname{Ran} D$ is dense in $\operatorname{Ran} D$ and $\operatorname{dom} D = \operatorname{Ker} D \oplus (\operatorname{dom} D \cap \operatorname{Ran} D)$, thus $D = 0 \oplus D|_{\operatorname{Ran} D}$ and $D|_{\operatorname{Ran} D}$ is invertible.*

PROOF. (1) Since, by Lemma 5.1.10, $D \in B(H(D), H)$ and since the range of D is closed, Prop. 5.1.2 implies that the range is complemented. Its complement is $\operatorname{Coker} D$. Since D is selfadjoint, $\operatorname{Ker} D = \operatorname{Coker} D$ by Lemma 5.1.9.

(2) Let $P : H \to \operatorname{Ran} D$ be the orthogonal projection. From $(1 - P)(\operatorname{dom} D) \subset \operatorname{Ker} D \subset \operatorname{dom} D$ we conclude that $P(\operatorname{dom} D) \subset \operatorname{dom} D$. The assertion follows because $P(\operatorname{dom} D)$ is dense in $P(H) = \operatorname{Ran} D$. □

We will need the following $\mathbb{Z}/2$-version of the previous proposition:

If $H = H^+ \oplus H^-$ is $\mathbb{Z}/2$-graded, then we call a closed operator D on H even resp. odd if $\operatorname{dom} D$ decomposes in $(\operatorname{dom} D)^+ \oplus (\operatorname{dom} D)^-$ and if the action of D is even resp. odd.

PROPOSITION 5.1.13. *Let H be a $\mathbb{Z}/2$-graded Hilbert $\mathcal{A}$-module and let D be an odd regular selfadjoint operator on H.*

Suppose that $D^+ : (\operatorname{dom} D)^+ \to H^-$ is surjective.

Then the range of D is closed and complemented. Furthermore $\operatorname{Ker} D^+ = \operatorname{Ker} D = \operatorname{Coker} D = \operatorname{Coker} D^-$ and this module is complemented.

PROOF. Since D^+ is surjective, D^- is injective and so $\operatorname{Ker} D^+ = \operatorname{Ker} D$.

Let P_+ be the orthogonal projection onto H^+. Since D is odd, $DP_+ = P_+ D$.

By Lemma 5.1.10 the operator $DP_+ : H(D) \to H$ is adjointable with adjoint $P_+ D(D^2 + 1)^{-1}$. From $P^+ D|_{H(D)^{\pm}} = D^{\pm}$ it follows that $D^-(D^2 + 1)^{-1} : H^- \to H(D)^+$ is the adjoint of $D^+ : H(D)^+ \to H^-$.

Since D^+ is surjective, $\operatorname{Ker} D^+$ is complemented in $H(D)^+$ and the range of the adjoint $D^-(D^2+1)^{-1}: H^- \to H(D)^+$ is closed. Furthermore
$$D^-(D^2+1)^{-1} = (D^2+1)^{-\frac{1}{2}} D^- (D^2+1)^{-\frac{1}{2}},$$
and $(D^2+1)^{-1/2}: H^\pm \to H(D)^\pm$ is an isomorphism by Lemma 5.1.10, hence $\operatorname{Ran} D^-$ is closed, too. □

PROPOSITION 5.1.14. *Let D be a regular selfadjoint operator on H.*
(1) *For all $\lambda \in \mathbb{C} \setminus \mathbb{R}$ the operator $D - \lambda$ is invertible.*
(2) *Assume that the range of D is closed and let P be the projection onto the kernel of D. Then there is $c > 0$ such that the spectrum of $(D+P)$ is contained in $\mathbb{R}\setminus]-c,c[$ and the spectrum of D is contained in $(\mathbb{R}\setminus]-c,c[) \cup \{0\}$.*

PROOF. This follows from the functional calculus for regular operators ([**La**], Th. 10.9) and from the decomposition in Prop. 5.1.12. □

The following criteria for selfadjointness are useful:

LEMMA 5.1.15. *Let D be a symmetric regular operator such that the range of $D + i$ and of $D - i$ is dense in H. Then D is selfadjoint.*

PROOF. The operators $D+i$ and of $D-i$ have closed range (see [**La**], Lemma 9.7). It follows that they have a bounded inverse on H. Then they are adjoint to each other, thus D is selfadjoint. □

LEMMA 5.1.16. *Assume that D is symmetric and has an inverse $D^{-1} \in B(H)$. Then D is regular.*

PROOF. Since D is symmetric, the adjoint is densely defined. From $D^{-1} \in B(H)$ it follows that the graph of D^{-1} is complemented, hence the graph of D is complemented as well. Hence D is regular. □

5.1.4. Decompositions of Hilbert C^*-modules. Let H be a Hilbert $\mathcal{A}$-module with $\mathcal{A}$-valued scalar product $\langle\,,\,\rangle$. Let $J = \{1, \ldots, m\} \subset \mathbb{N}$ resp. $J = \mathbb{N}$. If $J = \mathbb{N}$, then set $m = \infty$.

DEFINITION 5.1.17. *A system $\{f_k\}_{k \in J} \subset H$ is called* ORTHONORMAL *if for all $k, l \in J$*
$$\langle f_k, f_l \rangle = \delta_{kl}.$$
It is called an ORTHONORMAL BASIS OF H *if for all $f \in H$ there is $(a_n)_{n \in J} \subset \mathcal{A}$ such that $f = \sum_{n=1}^m f_n a_n$.*

Since $a_n = \langle f_n, f \rangle$, the coefficients are uniquely defined by the system.

PROPOSITION 5.1.18. *Let $\{f_k\}_{k \in J}$ be an orthonormal system in H whose span is dense in H. Then it is an orthonormal basis of H and the map*
$$f \mapsto (\langle f_n, f \rangle)_{n \in J}$$
is an isomorphism from H to $\mathcal{A}^m$ if $m < \infty$ and to $l^2(\mathcal{A})$ else.

PROOF. Let P_n be the orthogonal projection onto the span of the first n vectors of the system $\{f_k\}_{k\in J}$. On $\text{span}_\mathcal{A}\{f_k \mid k \in J\}$ the projection P_n converges strongly to the identity for $n \to \infty$. Since $\|P_n\| = 1$ for all $n \in \mathbb{N}$, it follows that P_n converges strongly to the identity on H. □

LEMMA 5.1.19. *Let $\{U_i\}_{i\in\mathbb{N}}$ be a family of pairwise orthogonal closed subspaces of H such that $\oplus_{i\in\mathbb{N}} U_i$ is dense in H. Let $\{T_i\}_{i\in\mathbb{N}}$ be a family of operators with $T_i \in B(U_i)$ and assume that there is $c \in \mathbb{R}$ such that $\|T_i\| \leq c$ for all $i \in \mathbb{N}$. Then the closure T of the operator $\oplus_{i\in\mathbb{N}} T_i$ is in $B(H)$ and $\|T\| \leq c$.*

PROOF. The spectral radius of an operator $A \in B(H)$ is denoted by $r(A)$. Write $T(n)$ for the restriction of T on $\oplus_{i=1}^n U_i$. Then for all $n \in \mathbb{N}$
$$\|T(n)\|^2 = r(T(n)^* T(n)) = \max_{1\leq i \leq n} r(T_i^* T_i) = \max_{1\leq i \leq n} \|T_i^* T_i\| \leq c^2 .$$
For $v \in \text{dom}\, T$ there is $n \in \mathbb{N}$ such that $v \in \oplus_{i=1}^n U_i$. Then $Tv = T(n)v$ and thus
$$\|Tv\| = \|T(n)v\| \leq c\|v\| .$$
It follows that the closure of $\oplus_{i\in\mathbb{N}} T_i$ is a bounded operator on H. The adjoint is given by the closure of $\oplus_{i\in\mathbb{N}} T_i^*$. □

COROLLARY 5.1.20. *Let $\{U_i\}_{i\in\mathbb{N}}$ be a family of pairwise orthogonal closed subspaces of H such that $\oplus_{i\in\mathbb{N}} U_i$ is dense in H. Let $\{T_i\}_{i\in\mathbb{N}}$ be a family of operators such that $T_i^{-1} \in B(U_i)$ and assume that there is $c \in \mathbb{R}$ such that $\|T_i^{-1}\| \leq c$ for all $i \in \mathbb{N}$. Then the closure T of the map $\oplus_{i\in\mathbb{N}} T_i$ is invertible with inverse in $B(H)$.*

PROOF. The operator $\oplus_{i\in\mathbb{N}} T_i^{-1}$ is inverse to $\oplus_{i\in\mathbb{N}} T_i$. It fulfills the conditions of the previous lemma, hence its closure is a bounded operator on H. It is the inverse of the closure of T. □

PROPOSITION 5.1.21. *Let $\{e_i\}_{i\in\mathbb{N}}$ be the standard basis of $l^2(\mathcal{A})$. Let M be a closed and N a projective submodule of $l^2(\mathcal{A})$ such that $l^2(\mathcal{A}) = M \oplus N$. Let P be the projection onto N along M and let P_n be the orthogonal projection onto $L_n := \text{span}_\mathcal{A}\{e_i \mid i = 1, \ldots, n\}$. Assume that P is adjointable.*
For $n \in \mathbb{N}$ such that
$$\|P(1 - P_n)\| \leq \frac{1}{2}$$
it holds:

(i) *The $\mathcal{A}$-module $N' := P_n(N)$ is projective and the maps*
$$P_n : N \to N' \text{ and } P : N' \to N$$
are isomorphisms.

(ii) $l^2(\mathcal{A}) = M \oplus N'$.

Note that for $n \in \mathbb{N}$ large enough $\|P(1-P_n)\| \leq \frac{1}{2}$ since P is a compact operator.

PROOF. From $\|P(1-P_n)\| \leq \frac{1}{2}$ it follows that
$$\|1_N - (PP_n)|_N\| \leq \frac{1}{2} .$$
Hence $(PP_n)|_N : N \to N$ is invertible.

The module $N' := P_n(N)$ is closed and finitely generated, thus projective. Furthermore, since $(PP_n)|_N$ is an isomorphism, the maps
$$P_n : N \to N' \text{ and } P : N' \to N$$
are isomorphisms as well.

It remains to show $N' \oplus M = l^2(\mathcal{A})$.

The intersection $N' \cap M$ is trivial: If $x \in N' \cap M$, then $Px = 0$, hence, as $P : N' \to N$ is an isomorphism, $x = 0$.

Let now $x \in l^2(\mathcal{A})$. Since $PP_n : N \to N$ is invertible, there is $y \in N$ such that $Px = PP_n y$.

Then
$$\begin{aligned}
x &= (1-P)x + Px \\
&= (1-P)x + PP_n y \\
&= (1-P)x + P_n y - (1-P)P_n y \\
&= (1-P)(x - P_n y) + P_n y \ .
\end{aligned}$$

Since $(1-P)(x - P_n y) \in M$ and $P_n y \in N'$, it follows that $x \in M \oplus N'$. $\square$

5.2. Operators on spaces of vector valued functions

5.2.1. Vector valued functions and tensor products. Let V be a Fréchet space.

In the following we are interested in spaces $\Gamma(M, V)$ where $\Gamma = C, C^\infty, C^k, L^p$ and M is endowed with the appropriate structure. It is desirable to have an isomorphism $\Gamma(M, \mathbb{C}) \otimes V \cong \Gamma(M, V)$ extending the inclusion $\Gamma(M, \mathbb{C}) \odot V \hookrightarrow \Gamma(M, V)$ because this ensures that every bounded operator on $\Gamma(M, \mathbb{C})$ extends to a bounded operator on $\Gamma(M, V)$ (at least if the tensor product is an ε- or π-tensor product). Examples where such an isomorphism exists are listed below whereas the example $\Gamma = L^2$ where in general such an isomorphism cannot be found is studied in detail in the following sections.

Proofs can be found in [**Tr**]. If not specified the functions are assumed to be complex valued.

- Let M be a compact topological space. Then
$$C(M) \otimes_\varepsilon V \cong C(M, V) \ .$$
 For compact spaces M, N
$$C(M \times N) \cong C(M) \otimes_\varepsilon C(N) \ .$$

- Let $U \subset \mathbb{R}^n$ be open and precompact. For all $m \in \mathbb{N}_0$
$$C_0^m(U) \otimes_\varepsilon V \cong C_0^m(U, V) \ .$$

- Let M be a closed smooth manifold. For all $m \in \mathbb{N}_0$
$$C^m(M) \otimes_\varepsilon V \cong C^m(M, V) \ .$$

- Let $U \subset \mathbb{R}^n$ be open and precompact. Then $C_0^\infty(U)$ is nuclear, in particular
$$C_0^\infty(U) \otimes_\pi V \cong C_0^\infty(U) \otimes_\varepsilon V \cong C_0^\infty(U, V) \ .$$

- Let M be a closed smooth manifold. Then $C^\infty(M)$ is nuclear, in particular
$$C^\infty(M) \otimes_\pi V \cong C^\infty(M) \otimes_\varepsilon V \cong C^\infty(M, V) \ .$$
For closed smooth manifolds M, N
$$C^\infty(M \times N) \cong C^\infty(M) \otimes C^\infty(N) \ .$$
- The space of Schwartz functions $\mathcal{S}(\mathbb{R})$ is nuclear, in particular
$$\mathcal{S}(\mathbb{R}) \otimes_\pi V \cong \mathcal{S}(\mathbb{R}) \otimes_\varepsilon V \cong \mathcal{S}(\mathbb{R}, V) \ .$$

The isomorphisms are given by the unique extension of the inclusion of the algebraic tensor product.

5.2.2. L^2-spaces and integral operators. Let E be a Banach space with norm $|\cdot|$. Let $\operatorname{End} E$ be the Banach algebra of bounded operators on E. We denote the operator norm on $\operatorname{End} E$ by $|\cdot|$ as well.

DEFINITION 5.2.1. *Let M be a measure space and $p \in \mathbb{N}$.*
The Banach space $L^p(M, E)$ is defined to be the completion of the quotient of the space of simple E-valued functions on M by the subspace of functions vanishing almost everywhere with respect to the norm
$$\|f\|_{L^p} := \left(\int_M |f(x)|^p dx \right)^{\frac{1}{p}} \ .$$

In order to avoid confusion we make the following convention: If $E = \mathcal{A}^n$ for a C^*-algebra $\mathcal{A}$, then $L^2(M, E)$ denotes the Hilbert $\mathcal{A}$-module defined in §5.1.1 and not the space just defined. In general these spaces do not coincide.

LEMMA 5.2.2. *Let M_1, M_2 be σ-finite measure spaces. Then the map*
$$L^2(M_1 \times M_2, E) \to L^2(M_1, L^2(M_2, E)), \ f \mapsto (x \mapsto f(x, \cdot))$$
is an isometric isomorphism.

PROOF. The lemma follows from Fubini. □

PROPOSITION 5.2.3. *Let M be a measure space.*
Let $k : M \times M \to \operatorname{End} E$ be a measurable function such that the integral kernel $|k(x, y)|$ defines a bounded operator $|K|$ on $L^2(M)$ with norm $|||K|||$. Then k defines a bounded operator on $L^2(M, E)$ with norm less than or equal to $|||K|||$.

PROOF. For a simple function $f : M \to E$
$$\| \int_M k(\cdot, y) f(y) \, dy \|_{L^2} \leq \| \int_M |k(\cdot, y)| |f(y)| \, dy \|_{L^2} \leq |||K||| \, \|f\|_{L^2}.$$
□

COROLLARY 5.2.4. *Let M be a measure space.*
There is a norm-decreasing map
$$L^2(M \times M, \operatorname{End} E) \to B(L^2(M, E), L^2(M, E))$$
$$k \mapsto \left(f \mapsto Kf := \int_M k(\cdot, y) f(y) \, dy \right) \ .$$

COROLLARY 5.2.5. *The convolution induces a continuous map*
$$L^1(\mathbb{R}^n, \operatorname{End} E) \to B(L^2(\mathbb{R}^n, E)), \ f \mapsto (g \mapsto f * g) \ .$$

PROOF. The convolution with $f \in L^1(\mathbb{R}^n, \operatorname{End} E)$ is an integral operator with integral kernel $f(x-y)$. Since $|f| \in L^1(\mathbb{R}^n)$ for $f \in L^1(\mathbb{R}^n, \operatorname{End} E)$, the convolution with $|f|$ is bounded on $L^2(\mathbb{R}^n)$. Then the assertion follows from the previous proposition. □

LEMMA 5.2.6. *For any $f \in L^p(\mathbb{R}^n, E)$ the translation map*

$$\tau f : \mathbb{R}^n \to L^p(\mathbb{R}^n, E), \ y \mapsto \tau_y f ,$$

defined by

$$\tau_y f(x) := f(x - y) ,$$

is continuous.

PROOF. The proof is analogous to the case $E = \mathbb{C}$, see [**Co**], Ch. VII, Prop. 9.2. □

5.2.3. Adjointable operators on Banach spaces. Let $\mathcal{B}$ be an involutive Banach algebra with unit. In this section all operators are assumed to be right $\mathcal{B}$-module maps. Let E be a Banach right $\mathcal{B}$-module with norm $|\cdot|$.

DEFINITION 5.2.7. *A $\mathcal{B}$-VALUED NON-DEGENERATED PRODUCT ON E is a map $\langle\, ,\, \rangle : E \times E \to \mathcal{B}$ that is $\mathbb{C}$-linear in the second variable and has the following properties:*
  (1) $\langle v, wb \rangle = \langle v, w \rangle b$ *for all $v, w \in E$, $b \in \mathcal{B}$,*
  (2) $\langle v, w \rangle = \langle w, v \rangle^*$ *for all $v, w \in E$,*
  (3) *if $\langle v, w \rangle = 0$ for all $w \in E$, then $v = 0$,*
  (4) *there is $C > 0$ such that $|\langle v, w \rangle| \leq C |v|_E \, |w|_E$ for all $v, w \in E$.*

Let E be endowed with a $\mathcal{B}$-valued non-degenerated product $\langle\, ,\, \rangle$.

DEFINITION 5.2.8. *A bounded operator $T : E \to E$ is called* ADJOINTABLE *if there is a map $T^* : E \to E$ satisfying*

$$\langle v, Tw \rangle = \langle T^* v, w \rangle$$

for all $v, w \in E$.

LEMMA 5.2.9. *Let $T : E \to E$ be adjointable.*
  (1) *The adjoint T^* is unique.*
  (2) *T^* is a right $\mathcal{B}$-module map.*
  (3) *T^* is bounded.*
  (4) *$T^{**} = T$.*
  (5) *$(ST)^* = T^* S^*$.*

PROOF. (1) Let T^* be an adjoint of T and $\Gamma(T^*)$ its graph. Then

$$\Gamma(T^*) \subset G := \{(x, y) \in E \times E \mid \langle y, w \rangle + \langle -x, Tw \rangle = 0 \ \forall w \in E\} .$$

Let $v \in E$. There is a unique $v_1 \in E$ with $(v, v_1) \in G$ since from

$$\langle v_1, w \rangle = \langle v, Tw \rangle = \langle v_2, w \rangle \ \forall w \in E$$

it follows $\langle v_1 - v_2, w \rangle = 0$ for all $w \in E$ and therefore $v_1 = v_2$. This shows $\Gamma(T^*) = G$.

(2) If $(x, y), (v, w) \in \Gamma(T^*)$ and $b \in \mathcal{B}$, then $(xb + v, yb + w) \in \Gamma(T^*)$ by the proof of (1).

(3) Since $\Gamma(T^*)$ is closed, the operator T^* is bounded.

(4) From (1) it follows that $\Gamma(T^{**}) = \Gamma(T)$.
(5) $\langle (ST)^*v, w\rangle = \langle v, STw\rangle = \langle S^*v, Tw\rangle = \langle T^*S^*v, w\rangle$. $\square$

LEMMA 5.2.10. *Let T be a densely defined operator on E. If there exists a densely defined operator S, called* FORMAL ADJOINT *of T, such that*
$$\langle f, Tg\rangle = \langle Sf, g\rangle$$
for all $f \in \operatorname{dom} S$, $g \in \operatorname{dom} T$, then T is closable.

PROOF. The set
$$\{(x,y) \in E \oplus E \mid \langle f, y\rangle - \langle Sf, x\rangle = 0 \ \forall f \in \operatorname{dom} S\}$$
is the graph of a closed extension of T. $\square$

The standard $\mathcal{B}$-valued non-degenerated product on $\mathcal{B}^n$ is given by
$$\langle v, w\rangle := \sum_{i=1}^{n} v_i^* w_i \ .$$
Since the endomorphism set of $\mathcal{B}^n$ can be identified with $M_n(\mathcal{B})$, all elements of $\operatorname{End}(\mathcal{B})$ are adjointable and taking the adjoint is a bounded linear map.

If M is a measure space, the standard $\mathcal{B}$-valued non-degenerated product on $L^2(M, \mathcal{B}^n)$ is defined by
$$\langle f, g\rangle_{L^2} := \int_M \langle f(x), g(x)\rangle dx \ .$$
We check condition (3) of Def. 5.2.7:

Let $\mathcal{B}'$ be the topological dual of $\mathcal{B}$ endowed with the weak topology.

We use that every $\lambda \in \mathcal{B}'$ induces a map $\lambda: \mathcal{B}^n \to \mathbb{C}^n$ by componentwise application.

If $f \in L^2(M, \mathcal{B}^n)$ with $\langle f, g\rangle_{L^2} = 0$ for all $g \in L^2(M, \mathcal{B}^n)$, then in particular for all $g \in L^2(M, \mathbb{C}^n)$ and $\lambda \in \mathcal{B}'$
$$\int_M \lambda(f(x)^*)g(x)dx = 0 \ ,$$
hence $\lambda(f(x)^*)$ vanishes almost everywhere. Since $\mathcal{B}'$ is separable, it follows $f = 0$ in $L^2(M, \mathcal{B}^n)$.

5.2.4. Hilbert-Schmidt operators. Let the notation be as in the previous section. Assume that M is a σ-finite measure space.

LEMMA 5.2.11. *Let $k \in L^2(M \times M, M_n(\mathcal{B}))$ and let K be the corresponding integral operator on $L^2(M, \mathcal{B}^n)$. Then k is uniquely defined by K.*

PROOF. It is enough to show that k vanishes in $L^2(M \times M, M_n(\mathcal{B}))$ if $K = 0$.

Applying $\lambda \in \mathcal{B}'$ componentwise yields maps $\lambda: M_n(\mathcal{B}) \to M_n(\mathbb{C})$ and $\lambda: \mathcal{B}^n \to \mathbb{C}^n$.

For $f \in L^2(M, \mathbb{C}^n)$ we have that almost everywhere
$$\lambda\left(\int_M k(x,y)f(y) \, dy\right) = \int_M \lambda(k(x,y))f(y) \, dy = 0 \ .$$
It follows that $\lambda(k(x,y)) = 0$ in $L^2(M \times M, M_n(\mathbb{C}))$. Since $\mathcal{B}'$ is separable, the set of all $(x,y) \in M \times M$ such that there is $\lambda \in \mathcal{B}'$ with $\lambda(k(x,y)) \neq 0$ has measure zero. On the complement of this set k vanishes. $\square$

DEFINITION 5.2.12. *A* HILBERT-SCHMIDT OPERATOR ON $L^2(M, \mathcal{B}^n)$ *is an integral operator with integral kernel in $L^2(M \times M, M_n(\mathcal{B}))$. Let A be a Hilbert-Schmidt operator on $L^2(M, \mathcal{B}^n)$ with integral kernel $k_A \in L^2(M \times M, M_n(\mathcal{B}))$. We define*

$$\|A\|_{HS} := \|k_A\|,$$

where the norm on the right hand side is taken in $L^2(M \times M, M_n(\mathcal{B}))$.

The normed space of Hilbert-Schmidt operators on $L^2(M, \mathcal{B}^n)$ is denoted by $HS(L^2(M, \mathcal{B}^n))$.

Note that $HS(L^2(M, \mathcal{B}^n))$ is a Banach algebra and that the inclusion

$$HS(L^2(M, \mathcal{B}^n)) \to B(L^2(M, \mathcal{B}^n))$$

is bounded. Prop. 5.2.13 below shows that $HS(L^2(M, \mathcal{B}^n))$ is a left $B(L^2(M, \mathcal{B}^n))$-module.

All operators in $HS(L^2(M, \mathcal{B}^n))$ are adjointable. The integral kernels of $A \in HS(L^2(M, \mathcal{B}^n))$ and A^* are related by $k_{A^*}(x, y) = k_A(y, x)^*$. It follows that taking the adjoint is a bounded map on $HS(L^2(M, \mathcal{B}^n))$.

PROPOSITION 5.2.13. *Let $A \in B(L^2(M, \mathcal{B}^n))$, $K \in HS(L^2(M, \mathcal{B}^n))$.*

(1) *Then $AK \in HS(L^2(M, \mathcal{B}^n))$. Furthermore there is $C > 0$, independent of A and K, such that*

$$\|AK\|_{HS} \le C \|A\| \|K\|_{HS}.$$

(2) *If A is adjointable, then $KA \in HS(L^2(M, \mathcal{B}^n))$. Furthermore there is $C > 0$, independent of A and K, with*

$$\|KA\|_{HS} \le C \|A^*\| \|K\|_{HS}.$$

PROOF. (1) There is an isomorphism

$$L^2(M \times M, M_n(\mathcal{B})) \cong L^2(M \times M, \mathcal{B}^n)^n$$

that is equivariant with respect to the left $M_n(\mathcal{B})$-action on both spaces. Furthermore the map

$$L^2(M \times M, \mathcal{B}^n) \to L^2(M, L^2(M, \mathcal{B}^n)), \; k \mapsto (y \mapsto k(\cdot, y))$$

is an isomorphism by Lemma 5.2.2. The operator A induces a bounded map on $L^2(M, L^2(M, \mathcal{B}^n))$, namely $k \mapsto (y \mapsto Ak(\cdot, y))$, clearly its norm is less than or equal to the norm of A on $L^2(M, \mathcal{B}^n)$.

(2) The map $K \mapsto KA$ is a composition of the following maps on $HS(L^2(M, \mathcal{B}^n))$:

$$K \stackrel{*}{\mapsto} K^* \stackrel{A^*}{\mapsto} A^*K^* = (KA)^* \stackrel{*}{\mapsto} KA.$$

By (1) and the fact that taking the adjoint is bounded on $HS(L^2(M, \mathcal{B}^n))$ these maps are bounded. □

DEFINITION 5.2.14. (1) *Let $\langle \, , \, \rangle : \mathcal{B}^n \times \mathcal{B}^n \to \mathcal{B}$ be the standard $\mathcal{B}$-valued non-degenerated product. For $e \in \mathcal{B}^n$ define the map*

$$e^* : \mathcal{B}^n \to \mathcal{B}, \; v \mapsto \langle e, v \rangle.$$

(2) *An integral operator A on $L^2(M, \mathcal{B}^n)$ is called* FINITE *if there is $k \in \mathbb{N}$ and there are functions $f_j, h_j \in L^2(M, \mathcal{B}^n)$, $j = 1 \ldots k$, such that*

$$k_A(x, y) = \sum_{j=1}^{k} f_j(x) h_j(y)^*.$$

5.2. OPERATORS ON SPACES OF VECTOR VALUED FUNCTIONS

5.2.5. Trace class operators. Let M be a σ-finite Borel space. Assume that there is a uniformly bounded sequence $(K_m)_{m \in \mathbb{N}} \subset B(L^2(M, \mathcal{B}^n))$ converging strongly to the identity such that each K_m is an integral operator with continuous compactly supported integral kernel.

This condition is fulfilled for example if M is a complete Riemannian manifold: Using Prop. 5.2.3 and the fact that the heat kernel $k_t(x, y)$ of the scalar Laplacian Δ is positive, one can deduce that $k_t(x, y)$ defines a strongly continuous semigroup $e^{-t\Delta}$ on $L^2(M, \mathcal{B}^n)$. Then $K_m := \phi_m e^{-\frac{1}{m}\Delta} \phi_m$ is an appropriate sequence, where $(\phi_m)_{m \in \mathbb{N}}$ is a sequence in $C_c(M)$ that converges to the identity on compact subsets of M.

The composition of operators induces a continuous map
$$\mu : HS(L^2(M, \mathcal{B}^n)) \otimes_\pi HS(L^2(M, \mathcal{B}^n)) \to B(L^2(M, \mathcal{B}^n)) .$$

DEFINITION 5.2.15. *A* TRACE CLASS OPERATOR *on $L^2(M, \mathcal{B}^n)$ is an element in the range of μ. The space of trace class operators is denoted by $\mathrm{Tr}(L^2(M, \mathcal{B}^n))$. We identify it with*
$$\left(HS(L^2(M, \mathcal{B}^n)) \otimes_\pi HS(L^2(M, \mathcal{B}^n))\right)/\mathrm{Ker}\,\mu$$
and endow it with the quotient norm.

It follows from Prop. 5.2.13 that $\mathrm{Tr}(L^2(M, \mathcal{B}^n))$ is a left $B(L^2(M, \mathcal{B}^n))$-module and that there is a well-defined action of adjointable operators from the right.

PROPOSITION 5.2.16. *The map*
$$R : HS(L^2(M, \mathcal{B}^n)) \otimes_\pi HS(L^2(M, \mathcal{B}^n)) \to L^1(M, M_n(\mathcal{B})) ,$$
$$(A, B) \to \left(x \mapsto \int_M k_A(x, y) k_B(y, x) \, dy\right)$$
descends to a continuous map
$$\overline{R} : \mathrm{Tr}(L^2(M, \mathcal{B}^n)) \to L^1(M, M_n(\mathcal{B})) .$$

PROOF. Note that R is continuous, hence all we have to show is that $\overline{R}$ is well-defined. Since the sequence K_m is uniformly bounded, the map $F \mapsto K_m F K_m$ on $HS(L^2(M, \mathcal{B}^n)) \otimes_\pi HS(L^2(M, \mathcal{B}^n))$ converges strongly to the identity. The assertion follows if we can show that $\mu(F) = 0$ implies $R(K_m F K_m) = 0$.

The map $F \mapsto K_m F K_m$ is continuous as a map from $HS(L^2(M, \mathcal{B}^n)) \otimes_\pi HS(L^2(M, \mathcal{B}^n))$ to $C_c(M, L^2(M, M_n(\mathcal{B}))) \otimes_\pi C_c(M, L^2(M, M_n(\mathcal{B})))$. The composition with the continuous map $C_c(M, L^2(M, M_n(\mathcal{B}))) \otimes_\pi C_c(M, L^2(M, M_n(\mathcal{B}))) \to C_c(M \times M, M_n(\mathcal{B}))$ induced by the product
$$L^2(M, M_n(\mathcal{B})) \times L^2(M, M_n(\mathcal{B})) \to M_n(\mathcal{B}), \ (f, g) \mapsto \int_M f(x) g(x) \, dx$$
agrees with the map that sends F to the continuous integral kernel $k_{\mu(K_m F K_m)}$ of $\mu(K_m F K_m)$. We compose further with the restriction to the diagonal and get a continuous map
$$HS(L^2(M, \mathcal{B}^n)) \otimes_\pi HS(L^2(M, \mathcal{B}^n)) \to C_c(M, M_n(\mathcal{B})) \subset L^1(M, M_n(\mathcal{B})) ,$$
which agrees with the map $F \mapsto R(K_m F K_m)$. This shows that the map $F \mapsto R(K_m F K_m)$ factors through the map
$$HS(L^2(M, \mathcal{B}^n)) \otimes_\pi HS(L^2(M, \mathcal{B}^n)) \to C_c(M \times M, M_n(\mathcal{B})), \ F \mapsto k_{\mu(K_m F K_m)} .$$

Now $\mu(F) = 0$ implies $\mu(K_m F K_m) = 0$, hence $k_{\mu(K_m F K_m)} = 0$ by the uniqueness of the integral kernel in $C_c(M \times M, M_n(\mathcal{B}))$. □

Let
$$\mathrm{tr} : M_n(\mathcal{B}) \to \mathcal{B}/\overline{[\mathcal{B},\mathcal{B}]}$$
be the trace defined, as usual, by adding up the diagonal elements and let
$$\mathrm{Tr}(T) := \int_M \mathrm{tr}\ \overline{R}(T)(x)\ dx$$
for $T \in \mathrm{Tr}(L^2(M, M_n(\mathcal{B})))$. Then for Hilbert-Schmidt operators A, B
$$\mathrm{Tr}(AB) = \mathrm{Tr}(BA)\ .$$

From the fact that any trace class operator can be approximated by finite sums of products of Hilbert-Schmidt operators it follows that the equation holds also for $A \in \mathrm{Tr}(L^2(M, \mathcal{B}^n))$ and B any adjointable operator.

5.2.6. Pseudodifferential operators. Let E be a Banach space.

Let U be an open precompact subset of $\mathbb{R}^n$. Recall the notion of a symbol of order m on U:

DEFINITION 5.2.17. *A function $a \in C^\infty(U \times \mathbb{R}^n, M_l(\mathbb{C}))$ is called a* SYMBOL OF ORDER *$m \in \mathbb{R}$ if it is compactly supported in the first variable and if for all multi-indices $\alpha, \beta \in \mathbb{N}_0^n$ the expressions*
$$\sup_{x \in U,\ \xi \in \mathbb{R}^n} (1 + |\xi|)^{-m+|\beta|} |\partial_x^\alpha \partial_\xi^\beta a(x,\xi)|$$
are finite.

These are seminorms on the space $S^m(U, M_l(\mathbb{C}))$ of symbols of order m on U, which turn $S^m(U, M_l(\mathbb{C}))$ into a Fréchet space.

In order to simplify formula involving Fourier transform it is convenient to rescale the Lebesgue measure on $\mathbb{R}^n$ by setting $d'x := (2\pi)^{-\frac{n}{2}} dx$.

We consider $L^2(U, E^l)$ as a subspace of $L^2(\mathbb{R}^n, E^l)$ in the following.

The Fourier transform is bounded from $L^1(\mathbb{R}^n, E)$ to $C_0(\mathbb{R}^n, E)$.

A symbol $a \in S^m(U, M_l(\mathbb{C}))$ defines a continuous operator
$$\mathrm{Op}(a) : C_c^\infty(U, E^l) \to C_c^\infty(U, E^l),\ (\mathrm{Op}(a)f)(x) = \int_{\mathbb{R}^n} e^{ix\xi} a(x,\xi) \hat{f}(\xi)\ d'\xi$$
called a pseudodifferential operator of order m.

Due to the fact that the Fourier transform is in general not continuous on $L^2(\mathbb{R}^n, E^l)$, the continuity properties of pseudodifferential operators acting on vector valued functions are in general weaker that those one gets for $E = \mathbb{C}$. They are still weaker if we allow for symbols with values in $\mathrm{End}\, E$.

LEMMA 5.2.18. *(1) For $m < -\frac{n}{2}$ and $a \in S^m(U, M_l(\mathbb{C}))$ the operator $\mathrm{Op}(a)$ extends to a bounded operator on $L^2(U, E^l)$ and the map*
$$\mathrm{Op} : S^m(U, M_l(\mathbb{C})) \to B(L^2(U, E^l))$$
is continuous.

(2) Let $m < -\frac{n}{2}$ and $\nu, k \in \mathbb{N}_0$ with $k < -\frac{n}{2} - m$. Then for $a \in S^m(U, M_l(\mathbb{C}))$ the operators
$$\mathrm{Op}(a): C_0^\nu(U, E^l) \to C_0^{\nu+k}(U, E^l)$$
and
$$\mathrm{Op}(a): L^2(U, E^l) \to C_0^k(U, E^l)$$
are continuous.

PROOF. (1) Let $m < -\frac{n}{2}$.
The Fourier transform in ξ induces a bounded map
$$S^m(U, M_l(\mathbb{C})) \to C_c^\infty(U, L^2(\mathbb{R}^n, M_l(\mathbb{C}))), \ a \mapsto (x \mapsto \hat{a}(x, \cdot)) \ .$$
If $a \in S^m(U, M_l(\mathbb{C}))$, we extend $\mathrm{Op}(a)$ to $L^2(U, E^l)$ by
$$(\mathrm{Op}(a)f)(x) := \int_{\mathbb{R}^n} \hat{a}(x, z) f(-x - z) d'z \ .$$
The map
$$S^m(U, M_l(\mathbb{C})) \to B(L^2(U, E^l)), \ a \mapsto \mathrm{Op}(a)$$
is well-defined and continuous since
$$\mathbb{R}^n \to L^2(\mathbb{R}^n, E^l), \ x \mapsto (z \mapsto f(-x - z))$$
is continuous by Lemma 5.2.6, hence $\mathrm{Op}(a)f \in C_0(U, E^l)$ and
$$\|\mathrm{Op}(a)f\|_{C_0} \le \sup_{x \in U} \|\hat{a}(x, \cdot)\|_{L^2} \|f\|_{L^2} \ .$$

(2) First let $m < -\frac{n}{2}$ and $k = 0$.
If $f \in C_0^\nu(U, E^l)$, then $x \mapsto (z \mapsto f(-x - z))$ is in $C_0^\nu(\mathbb{R}^n, L^2(\mathbb{R}^n, E^l))$.
It follows as above that $\mathrm{Op}(a)f \in C_0^\nu(U, E^l)$ and
$$\|\mathrm{Op}(a)f\|_{C^\nu} \le C \sup_{|\alpha| \le \nu} \sup_{x \in U} \|\partial_x^\alpha \hat{a}(x, \cdot)\|_{L^2} \|f\|_{C^\nu} \ .$$
Now assume that the assertion holds for $k - 1$ and all $a \in S^m(U, M_l(\mathbb{C}))$ with $m < -\frac{n}{2} - k + 1$.
We prove the assertion for k and $a \in S^m(U, M_l(\mathbb{C}))$ with $m < -\frac{n}{2} - k$:
If $\alpha \in \mathbb{N}_0^n$ with $|\alpha| = 1$, then the map
$$f \mapsto \partial^\alpha(\mathrm{Op}(a)f)$$
is a pseudodifferential operator of degree $m + 1$. By induction it is a bounded operator from $C^\nu(U, E^l)$ to $C^{\nu+k-1}(U, E^l)$. It follows that $\mathrm{Op}(a)$ is continuous from $C^\nu(U, E^l)$ to $C^{\nu+k}(U, E^l)$.

An analogous induction argument shows that $\mathrm{Op}(a)$ is continuous from $L^2(U, E^l)$ to $C_0^k(U, E^l)$. For $k = 0$ this was proved in (1). $\square$

5.3. Projective systems and function spaces

The projective systems $(\mathcal{A}_i)_{i \in \mathbb{N}_0}$ and $(\hat{\Omega}_{\le \mu} \mathcal{A}_i)_{i,\mu \in \mathbb{N}_0}$ from §1.3.3 induce projective systems of spaces $\left(L^2(M, \mathcal{A}_i^l)\right)_{i \in \mathbb{N}_0}$ and $\left(L^2(M, (\hat{\Omega}_{\le \mu} \mathcal{A}_i)^l)\right)_{i,\mu \in \mathbb{N}_0}$.

Recall the convention fixed in §5.2.2: The space $L^2(M, \mathcal{A}^l)$ is the Hilbert $\mathcal{A}$-module defined in §5.1.1. For $\mu \in \mathbb{N}_0$ and $i \in \mathbb{N}$ the space $L^2(M, (\hat{\Omega}_{\le \mu} \mathcal{A}_i)^l)$ was defined in §5.2.2.

In the following we investigate the behavior of some particular classes of operators on $L^2(M, (\hat{\Omega}_{\le \mu} \mathcal{A}_i)^l)$ under the projective limit.

5.3.1. Integral operators.
Hilbert-Schmidt operators on $L^2(M, \mathcal{A}_i^l)$ have the property that they extend to bounded operators on $L^2(M, (\hat{\Omega}_{\leq \mu}\mathcal{A}_j)^l)$ for all $j \in \mathbb{N}_0$ with $j \leq i$ and all $\mu \in \mathbb{N}_0$. We investigate how the spectrum depends on j, μ.

We extend a method developed by Lott ([**Lo3**], §6.1.) for closed manifolds to certain non-compact manifolds with boundary (in particular for the manifold defined in §1.1).

Let $[0,1]^n$ be endowed with a measure of the form hdx where h is a positive continuous function on $[0,1]^n$ and dx is the Lebesgue measure.

In the proof of the following proposition we use that there exists a Schauder basis of $C([0,1]^n)$ that is orthonormal in $L^2([0,1]^n)$ (here and in the following we understand $L^2([0,1]^n)$ with respect to hdx). For $h = 1$ a Franklin system [**Se**] yields such a basis $\{f_n\}_{n \in \mathbb{N}}$, then for general h the system $\{h^{-\frac{1}{2}} f_n\}_{n \in \mathbb{N}}$ is one.

The proposition still holds true if $[0,1]^n$ is replaced by a compact Borel space for which such a basis exists.

PROPOSITION 5.3.1. (1) *Let $[0,1]^n$ be endowed with a measure hdx as above. Let $k \in C([0,1]^n \times [0,1]^n, M_l(\mathcal{A}_i))$ and let K be the corresponding integral operator. Assume that $1 - K$ is invertible in $B(L^2([0,1]^n, \mathcal{A}^l))$. Then $1 - K : L^2([0,1]^n, (\hat{\Omega}_{\leq \mu}\mathcal{A}_i)^l) \to L^2([0,1]^n, (\hat{\Omega}_{\leq \mu}\mathcal{A}_i)^l)$ is invertible.*

(2) *Let M be a Riemannian manifold of dimension n, possibly with boundary. Suppose that there is a covering $\{K_m\}_{m \in \mathbb{N}}$ of M with K_m compact, $K_m \subset K_{m+1}$ and such that K_m is diffeomorphic to $[0,1]^n$ for every $m \in \mathbb{N}$. Let $k \in L^2(M \times M, M_l(\mathcal{A}_i)) \cap C(M \times M, M_l(\mathcal{A}_i))$ and assume furthermore that $x \mapsto k(x, \cdot)$ and $y \mapsto k(\cdot, y)$ are in $C(M, L^2(M, M_l(\mathcal{A}_i)))$.*

If $1 - K$ is invertible in $B(L^2(M, \mathcal{A}^l))$, then $1 - K$ is invertible in $B(L^2(M, (\hat{\Omega}_{\leq \mu}\mathcal{A}_i)^l))$.

PROOF. (1) Choose a basis of $C([0,1]^n)$ which is orthonormal with respect to hdx and let P_N denote the projection onto the first N basis vectors. The integral kernel of P_N is in $L^2([0,1]^n \times [0,1]^n)$, hence P_N acts continuously on $L^2([0,1]^n, (\hat{\Omega}_{\leq \mu}\mathcal{A}_i)^l)$.

We decompose $L^2([0,1]^n, (\hat{\Omega}_{\leq \mu}\mathcal{A}_i)^l)$ into the direct sum
$$P_N L^2([0,1]^n, (\hat{\Omega}_{\leq \mu}\mathcal{A}_i)^l) \oplus (1 - P_N) L^2([0,1]^n, (\hat{\Omega}_{\leq \mu}\mathcal{A}_i)^l)$$
and write
$$1 - K = \begin{pmatrix} a & b \\ c & d \end{pmatrix}$$
with respect to the decomposition.

If we find N such that d is invertible on $(1 - P_N)L^2([0,1]^n, (\hat{\Omega}_{\leq \mu}\mathcal{A}_i)^l)$ and prove that then $a - bd^{-1}c$ is invertible on $P_N L^2([0,1]^n, (\hat{\Omega}_{\leq \mu}\mathcal{A}_i)^l)$, we can conclude that $(1 - K)$ is invertible by the equality
$$\begin{pmatrix} a & b \\ c & d \end{pmatrix} = \begin{pmatrix} 1 & bd^{-1} \\ 0 & 1 \end{pmatrix} \begin{pmatrix} a - bd^{-1}c & 0 \\ 0 & d \end{pmatrix} \begin{pmatrix} 1 & 0 \\ d^{-1}c & 1 \end{pmatrix}.$$

First we show that
$$d = (1 - P_N)(1 - K)(1 - P_N)$$
is invertible for N big enough. By Prop. 5.2.13 the operator $(1 - P_N)K(1 - P_N)$ is a Hilbert-Schmidt operator. Its integral kernel is continuous.

For $N \to \infty$ the projections P_N converge strongly to the identity on $C([0,1]^n)$. Since
$$C([0,1]^n \times [0,1]^n, M_l(\mathcal{A}_i)) \cong C([0,1]^n) \otimes_\varepsilon C([0,1]^n) \otimes_\varepsilon M_l(\mathcal{A}_i) ,$$
there is N such that the norm of the integral kernel of $(1-P_N)K(1-P_N)$ is smaller than $\frac{1}{2}$ in $C([0,1]^n \times [0,1]^n, M_l(\mathcal{A}_i))$.

For that N the series
$$(1-P_N) + \sum_{\nu=1}^{\infty} ((1-P_N)K(1-P_N))^\nu$$
converges as a bounded operator on $(1-P_N)L^2([0,1]^n, (\hat{\Omega}_{\leq \mu}\mathcal{A}_i)^l)$ and inverts d. Hence $a - bd^{-1}c$ is well-defined.

Via the basis we identify $a - bd^{-1}c$ with an element of $M_{Nl}(\mathcal{A}_i)$. Since $1-K$ is invertible on $L^2(M, \mathcal{A}^l)$, the matrix $a - bd^{-1}c$ is invertible in $M_{Nl}(\mathcal{A})$. By Prop. 1.3.4 it follows that $a - bd^{-1}c$ is invertible in $M_{Nl}(\mathcal{A}_i)$ as well.

(2) Let $m \in \mathbb{N}$ be such that in $L^2(M \times M, M_l(\mathcal{A}_i))$
$$\|(1 - 1_{K_m}(x))k(x,y)(1 - 1_{K_m}(y))\| \leq \frac{1}{2} .$$

Write
$$(1-K) = \begin{pmatrix} a & b \\ c & d \end{pmatrix}$$
with respect to the decomposition
$$L^2(M, (\hat{\Omega}_{\leq \mu}\mathcal{A}_i)^l) = 1_{K_m} L^2(M, (\hat{\Omega}_{\leq \mu}\mathcal{A}_i)^l) \oplus (1 - 1_{K_m}) L^2(M, (\hat{\Omega}_{\leq \mu}\mathcal{A}_i)^l) .$$

By the choice of $m \in \mathbb{N}$ the entry $d = 1 - (1 - 1_{K_m})K(1 - 1_{K_m})$ is invertible on $(1 - 1_{K_m})L^2(M, (\hat{\Omega}_{\leq \mu}\mathcal{A}_i)^l)$.

We prove that $a - bd^{-1}c$ is invertible on $L^2(K_m, (\hat{\Omega}_{\leq \mu}\mathcal{A}_i)^l)$ and then the assertion follows as in the proof of (1).

On $L^2(K_m, (\hat{\Omega}_{\leq \mu}\mathcal{A}_i)^l)$
$$a - bd^{-1}c = 1_{K_m} - (1_{K_m} K 1_{K_m} + bd^{-1}c) .$$

The operator $1_{K_m} K 1_{K_m} + bd^{-1}c$ is an integral operator on $L^2(K_m, (\hat{\Omega}_{\leq \mu}\mathcal{A}_i)^l)$ with continuous integral kernel: The integral kernel of b is $1_{K_m}(x)k(x,y)(1 - 1_{K_m}(y))$ and $x \mapsto 1_{K_m}(x)k(x,\cdot)(1 - 1_{K_m})$ is in $C(K_m, L^2(M, M_l(\mathcal{A}_i)))$. For the integral kernel $(1 - 1_{K_m}(x))k(x,y)1_{K_m}(y)$ of c we have $y \mapsto (1 - 1_{K_m})k(\cdot,y)1_{K_m}(y) \in C(K_m, L^2(M, M_l(\mathcal{A}_i)))$. It follows that $bd^{-1}c$ is an integral operator with continuous kernel on $K_m \times K_m$. Clearly the integral kernel of $1_{K_m} K 1_{K_m}$ is continuous as well.

Since $a - bd^{-1}c$ is invertible on $L^2(K_m, \mathcal{A}^l)$ and since the measure on K_m pulled back by an orientation preserving diffeomorphism $[0,1]^n \to K_m$ is of the form hdx, we conclude by (1) that $a - bd^{-1}c$ is invertible on $L^2(K_m, (\hat{\Omega}_{\leq \mu}\mathcal{A}_i)^l)$ as well. $\square$

COROLLARY 5.3.2. *Let k be an integral kernel as in part (2) of the proposition and let K be the corresponding integral operator. Then for $\lambda \in \mathbb{C}^*$ the operator $K - \lambda$ is invertible on $L^2(M, (\hat{\Omega}_{\leq \mu}\mathcal{A}_i)^l)$ if $K - \lambda$ is invertible on $L^2(M, \mathcal{A}^l)$.*

PROOF. For $\lambda \in \mathbb{C} \setminus \{0\}$ the integral kernel k/λ fulfills the conditions of the lemma. Thus if $\lambda - K = \lambda(1 - K/\lambda)$ is invertible on $L^2(M, \mathcal{A}^l)$, then $\lambda - K$ is invertible on $L^2(M, (\hat{\Omega}_{\leq \mu}\mathcal{A}_i)^l)$ as well. $\square$

5.3.2. The Chern character.

DEFINITION 5.3.3. *Let V be a $\mathbb{Z}/2$-graded finite dimensional vector space. Let K be an integral operator on $L^2(M, V \otimes \hat{\Omega}_{\leq \mu} \mathcal{A}_i)$ with integral kernel $k : M \times M \to \text{End}(V) \otimes \hat{\Omega}_{\leq \mu} \mathcal{A}_i$. Then we define $\mathrm{d}(K)$ to be the integral operator on $L^2(M, V \otimes \hat{\Omega}_{\leq \mu} \mathcal{A}_i)$ with integral kernel $\mathrm{d}(k(x,y))$. (The action of d on $\text{End}(V) \otimes \hat{\Omega}_{\leq \mu} \mathcal{A}_i$ was described in §1.3.2.)*

Note that if K is of degree n with respect to the $\mathbb{Z}/2$-grading on $L^2(M, V \otimes \hat{\Omega}_{\leq \mu} \mathcal{A}_i)$, then
$$\mathrm{d}(Kf) = \mathrm{d}(K)f + (-1)^n K \, \mathrm{d} f \ .$$

LEMMA 5.3.4. *Let M be a σ-finite measure space.*
Let P be a Hilbert-Schmidt operator on $L^2(M, (\hat{\Omega}_{\leq \mu} \mathcal{A}_i)^l)$ with integral kernel in $L^2(M \times M, M_l(\mathcal{A}_i))$ and assume that $P^2 = P$. Then
$$P \, \mathrm{d} P \, \mathrm{d} P = P(\mathrm{d}(P))^2 \ .$$

PROOF. As for matrices (see the beginning of the proof of Prop. 1.3.3) we have that $(\mathrm{d} P) = P(\mathrm{d} P) + (\mathrm{d} P)P$ and $P(\mathrm{d} P)P = 0$.

It follows that
$$\begin{aligned} P \, \mathrm{d} P \, \mathrm{d} P &= P(\mathrm{d} P) \, \mathrm{d} P = P(\mathrm{d} P)^2 P \\ &= P(\mathrm{d} P)(\mathrm{d} P) - P(\mathrm{d} P)P(\mathrm{d} P) \\ &= P(\mathrm{d}(P))^2. \end{aligned}$$
$\square$

LEMMA 5.3.5. *Let $P : [0,1] \to HS(L^2(M, (\hat{\Omega}_{\leq \mu} \mathcal{A}_i)^l))$ be a differentiable path of Hilbert-Schmidt operators with integral kernels in $L^2(M \times M, M_l(\mathcal{A}_i))$, and assume that $P(t)^2 = P(t)$ for every $t \in [0,1]$.*
Then for $k \in \mathbb{N}_0$ in $\hat{\Omega}_{\leq \mu} \mathcal{A}_i / [\overline{\hat{\Omega}_{\leq \mu} \mathcal{A}_i, \hat{\Omega}_{\leq \mu} \mathcal{A}_i}]$
$$\operatorname{Tr} P(1)(\mathrm{d} P(1))^{2k} - \operatorname{Tr} P(0)(\mathrm{d} P(0))^{2k}$$
is exact.

PROOF. As for matrices (see Prop. 1.3.3). $\square$

If T is a trace class operator on $L^2(M, (\hat{\Omega}_{\leq \mu} \mathcal{A}_i)^l)$ restricting to a trace class operator on $L^2(M, (\hat{\Omega}_{\leq \nu} \mathcal{A}_j)^l)$ for all $j, \nu \in \mathbb{N}$ with $j \geq i$ and $\nu \geq \mu$, then by taking the projective limit we can consider $\operatorname{Tr}(T)$ as in element in $\hat{\Omega}_* \mathcal{A}_\infty / [\overline{\hat{\Omega}_* \mathcal{A}_\infty, \hat{\Omega}_* \mathcal{A}_\infty}]_s$. In the next lemma we show that under certain conditions the formula for the Chern character can be generalized to projections which are Hilbert-Schmidt operators.

PROPOSITION 5.3.6. *Let M be a Riemannian manifold of dimension d, possibly with boundary. Suppose that there is a covering $\{K_m\}_{m \in \mathbb{N}}$ of M with K_m compact, $K_m \subset K_{m+1}$ and such that K_m is diffeomorphic to $[0,1]^d$ for all $m \in \mathbb{N}$.*
Let $P \in B(L^2(M, \mathcal{A}^l))$ be a projection onto a projective submodule of $L^2(M, \mathcal{A}^l)$. Assume further that for any $i \in \mathbb{N}$ it restricts to a bounded projection on $L^2(M, \mathcal{A}_i^l)$ and that $P(L^2(M, \mathcal{A}_i^l)) \subset C(M, \mathcal{A}_i^l)$. Let
$$\operatorname{Ran}_\infty P := \bigcap_{i \in \mathbb{N}} P(L^2(M, \mathcal{A}_i^l)) \ .$$

(1) *The projection P is a Hilbert-Schmidt operator with integral kernel of the form $\sum_{j=1}^{m} f_j(x)h_j(y)^*$ with $f_j, h_j \in \operatorname{Ran}_\infty P$.*
(2) *The intersection $\operatorname{Ran}_\infty P$ is a projective $\mathcal{A}_\infty$-module. The classes $[\operatorname{Ran} P] \in K_0(\mathcal{A})$ and $[\operatorname{Ran}_\infty P] \in K_0(\mathcal{A}_\infty)$ correspond to each other under the canonical isomorphism $K_0(\mathcal{A}) \cong K_0(\mathcal{A}_\infty)$.*
(3) *In $H_*^{dR}(\mathcal{A}_\infty)$*

$$\operatorname{ch}[\operatorname{Ran}_\infty P] = \sum_{n=0}^{\infty} (-1)^n \frac{1}{n!} \operatorname{Tr}(P \, d \, P)^{2n} \ .$$

PROOF. (1) Let $\{e_n\}_{n \in \mathbb{N}} \subset C_0^\infty(M, \mathbb{C}^l)$ be an orthonormal basis of $L^2(M, \mathbb{C}^l)$. The orthogonal projection P_n onto the span of the first n basis vectors is a Hilbert-Schmidt operator with integral kernel in $C_0^\infty(M \times M, M_l(\mathbb{C}))$.
In particular $P_n \in B(L^2(M, \mathcal{A}_i^l))$ for any $i \in \mathbb{N}$.
First we consider the situation on $L^2(M, \mathcal{A}^l)$:
Since P is compact, there is $n \in \mathbb{N}$ such that on $L^2(M, \mathcal{A}^l)$

$$\|P(P_n - 1)\| \leq \frac{1}{2} \ .$$

Then by Prop. 5.1.21 the map $PP_nP : \operatorname{Ran} P \to \operatorname{Ran} P$ is an isomorphism. It follows that $\operatorname{Ker} PP_nP = (\operatorname{Ran} P)^\perp = \operatorname{Ker} P$ and therefore

$$P = 1 - P_{\operatorname{Ker} PP_nP} \ .$$

Here $P_{\operatorname{Ker} PP_nP}$ denotes the orthogonal projection onto $\operatorname{Ker} PP_nP$.
We can find $r > 0$ such that $B_r(0) \setminus \{0\}$ is in the resolvent set of PP_nP. Then

$$\begin{aligned} P &= 1 - P_{\operatorname{Ker} PP_nP} \\ &= 1 - \frac{1}{2\pi i} \int_{|\lambda|=r} (\lambda - PP_nP)^{-1} d\lambda \\ &= \frac{1}{2\pi i} \int_{|\lambda|=r} \left(\lambda^{-1} - (\lambda - PP_nP)^{-1}\right) d\lambda \\ &= -\frac{1}{2\pi i} PP_nP \int_{|\lambda|=r} \lambda^{-1}(\lambda - PP_nP)^{-1} d\lambda \ . \end{aligned}$$

The integral kernel of PP_nP fulfills the conditions of Prop. 5.3.1. From Cor. 5.3.2 we conclude that the spectrum of PP_nP in $B(L^2(M, \mathcal{A}_i^l))$ is independent of $i \in \mathbb{N}_0$, thus

$$R := \int_{|\lambda|=r} \lambda^{-1}(\lambda - PP_nP)^{-1} d\lambda$$

is a bounded operator on $L^2(M, \mathcal{A}_i^l)$ for all $i \in \mathbb{N}_0$.
This and the equation $PR = PRP$ show that P is an integral operator with integral kernel

$$k_P(x,y) = -\frac{1}{2\pi i} \sum_{j=1}^{n} Pe_j(x)(PR^* Pe_j(y))^* \ .$$

The integral kernel is in $L^2(M \times M, M_l(\mathcal{A}_i))$ for all $i \in \mathbb{N}_0$ and is of the form we asserted.
(2) Let P_n as above with $\|(P_n - 1)P\| \leq \frac{1}{2}$ in $B(L^2(M, \mathcal{A}^l))$.

The operator $(1 + (P_n - 1)P)$ is invertible in $B(L^2(M, \mathcal{A}^l))$, hence by Prop. 5.3.1 it is invertible in $B(L^2(M, \mathcal{A}^l_i))$ for any $i \in \mathbb{N}_0$. From $P_n P = (1+(P_n-1)P)P$ it follows that

$$P_n P(L^2(M, \mathcal{A}^l_i)) \cong \operatorname{Ran} P(L^2(M, \mathcal{A}^l_i))$$

for all $i \in \mathbb{N}_0$. Furthermore $Q := (1+(P_n-1)P)P(1+(P_n-1)P)^{-1} \in B(L^2(M, \mathcal{A}^l_i))$ is a projection onto $P_n P(L^2(M, \mathcal{A}^l_i))$. Identify $P_n(L^2(M, \mathcal{A}^l_i))$ with $\mathcal{A}^n_i$ via the basis and let $Q' \in M_n(\mathcal{A}_\infty)$ be the restriction of Q to $P_n(L^2(M, \mathcal{A}^l_i))$. From $\operatorname{Ran}_\infty P \cong Q'(\mathcal{A}^n_\infty)$ it follows that $[\operatorname{Ran} P] = [Q']$ in $K_0(\mathcal{A})$ and $[\operatorname{Ran}_\infty P] = [Q']$ in $K_0(\mathcal{A}_\infty)$. This shows the assertion.

(3) Let P_n, Q and Q' be as in the proof of (2). In $H^{dR}_*(\mathcal{A}_\infty)$

$$\begin{aligned}
\operatorname{ch}[\operatorname{Ran}_\infty P] &= \operatorname{ch}(Q') \\
&= \sum_{k=0}^{\infty} (-1)^k \frac{1}{k!} \operatorname{tr}(Q' \, d \, Q')^{2k} \\
&= \sum_{k=0}^{\infty} (-1)^k \frac{1}{k!} \operatorname{Tr}((P_n Q P_n) \, d(P_n Q P_n))^{2k} \\
&= \sum_{k=0}^{\infty} (-1)^k \frac{1}{k!} \operatorname{Tr}(P_n Q)(d \, P_n Q)^{2k} \\
&= \sum_{k=0}^{\infty} (-1)^k \frac{1}{k!} \operatorname{Tr}(Q(d \, Q)^{2k}) \, .
\end{aligned}$$

Since

$$H : [0,1] \to B(L^2(M, (\hat{\Omega}_{\le \mu} \mathcal{A}_i)^l)) \, ,$$
$$H(t) = (1 + t(P_n - 1)P)P(1 + t(P_n - 1)P)^{-1}$$

is a differentiable path of finite projections with $H(0) = P$ and $H(1) = Q$, the difference $\operatorname{Tr}(Q \, d \, Q)^{2k} - \operatorname{Tr}(P \, d \, P)^{2k}$ is exact in $\hat{\Omega}_* \mathcal{A}_\infty / [\hat{\Omega}_* \mathcal{A}_\infty, \hat{\Omega}_* \mathcal{A}_\infty]_s$ by the previous lemma.

This shows the assertion. $\square$

5.4. Holomorphic semigroups

5.4.1. General facts. Let X be a Banach space.

In order to fix notation we collect some well-known facts about holomorphic semigroups on X. A general reference is [**RR**].

For $\delta \in]0, \pi]$ let $\Sigma_\delta := \{\lambda \in \mathbb{C}^* \mid |\arg \lambda| < \delta\}$.

Recall that a family $S : \Sigma_\delta \cup \{0\} \to B(X)$ is called a holomorphic semigroup if it is a semigroup, i.e. $S(0) = 1$ and $S(t + s) = S(t)S(s)$ for all $s, t \in \Sigma_\delta \cup \{0\}$, moreover if it is strongly continuous on $\Sigma_{\delta'} \cup \{0\}$ for all $\delta' < \delta$ and holomorphic on Σ_δ.

Given a holomorphic semigroup $S(t)$, the unbounded operator Z with $Zx := \frac{d}{dt}(S(t)x)|_{t=0}$ whenever defined is called the generator of Z. It is densely defined and closed. Any subset of its domain that is dense in X and invariant under the action of the semigroup is a core for Z.

5.4. HOLOMORPHIC SEMIGROUPS

Recall that a densely defined operator Z on X generates a holomorphic semigroup e^{tZ} if and only if there is $\omega \in \mathbb{R}$ such that $Z + \omega$ is δ-sectorial for some $\delta \in]0, \pi/2]$, i.e. if and only if $\Sigma_{\delta+\pi/2}$ is a subset of the resolvent set $\rho(Z + \omega)$ and for any ε with $0 < \varepsilon < \delta$ there is $C > 0$ such that for all $\lambda \in \Sigma_{\varepsilon+\pi/2}$

$$\|(Z + \omega - \lambda)^{-1}\| \leq \frac{C}{|\lambda|} .$$

Then there is $C > 0$ such that for all $t > 0$

$$\|e^{tZ}\| \leq Ce^{-\omega t} .$$

LEMMA 5.4.1. *Let Z be a densely defined operator on X and let $\delta > 0$ be such that*

(i) $\Sigma_{\delta+\pi/2} \cup \{0\} \subset \rho(Z)$,
(ii) *for every $\alpha < \delta$ there are $c \in \mathbb{R}$ and $r > 0$ such that*

$$\|(Z + c - \lambda)^{-1}\| \leq \frac{C}{|\lambda|}$$

for $\lambda \in \Sigma_{\alpha+\pi/2}$ with $|\lambda| > r$ and $\lambda - c \in \rho(Z)$.

Then Z is δ-sectorial.

PROOF. Let $0 < \varepsilon < \delta$.

Let $\alpha \in (\varepsilon, \delta)$ and $R > r$ big enough such that any $\lambda \in \Sigma_{\varepsilon+\pi/2}$ with $|\lambda| > R$ satisfies $\lambda - c \in \Sigma_{\alpha+\pi/2}$. By assumption we find $C > 0$ such that for all $\lambda \in \Sigma_{\varepsilon+\pi/2}$ with $|\lambda| > R$

$$\begin{aligned}
\|(Z - \lambda)^{-1}\| &\leq \frac{C}{|\lambda + c|} \\
&\leq \left(\frac{C}{|\lambda|}\right)\left(\frac{|\lambda + c| + |c|}{|\lambda + c|}\right) \\
&\leq \left(\frac{C}{|\lambda|}\right)\left(1 + \frac{|c|}{(R - |c|)}\right) \\
&\leq \frac{C}{|\lambda|} .
\end{aligned}$$

Since $\overline{\Sigma_{\varepsilon+\pi/2}} \cap \{|z| \leq R\}$ is a compact subset of the resolvent set of Z, this implies the assertion. □

We have the following relation between the spectrum of a δ-sectorial operator Z and the behavior of the holomorphic semigroup e^{tZ} for $t \to \infty$.

PROPOSITION 5.4.2. *If Z is a δ-sectorial operator and there is $\omega > 0$ such that*

$$\{\operatorname{Re}\lambda > -\omega\} \subset \rho(Z) ,$$

then for any $\omega' < \omega$ there is $C > 0$ such that for all $t \geq 0$

$$\|e^{tZ}\| \leq Ce^{-\omega' t} .$$

PROOF. For any $\omega' < \omega$ there is $0 < \delta' \leq \delta$ such that $\Sigma_{\delta'+\pi/2} \cup \{0\} \subset \rho(Z+\omega')$. Now it follows from the previous lemma that $Z + \omega'$ is δ'-sectorial. □

PROPOSITION 5.4.3. *If e^{tZ} is a strongly continuous semigroup with generator Z such that $\operatorname{Ran} e^{tZ} \subset \operatorname{dom} Z$ for all $t \in (0, \infty)$ and if there are $T > 0$ and $C > 0$ such that for $t \leq T$*
$$\|Ze^{tZ}\| \leq Ct^{-1},$$
then e^{tZ} extends to a holomorphic semigroup.

If the estimate holds for all $t > 0$, then the extension is bounded holomorphic.

PROOF. The assertion follows immediately from [**Da**], Th. 2.39. □

Part (1) of the next proposition is known under the name Volterra development and the formula in part (2) is called Duhamel's formula:

PROPOSITION 5.4.4. *Let Z be the generator of a strongly continuous semigroup and let $M, \omega \in \mathbb{R}$ be such that $\|e^{tZ}\| \leq Me^{\omega t}$ for all $t \geq 0$.*

(1) *Let $R \in B(X)$. Then $Z + R$ is the generator of a strongly continuous semigroup and for all $t \geq 0$*
$$e^{t(Z+R)} = \sum_{n=0}^{\infty} t^n \int_{\Delta^n} e^{u_0 tZ} R e^{u_1 tZ} R \ldots e^{u_n tZ} \, du_0 \ldots du_n$$
with $\Delta^n = \{u_0 + \cdots + u_n = 1; u_i \geq 0, \ i = 0, \ldots, n\}$. Furthermore
$$\|e^{t(Z+R)}\| \leq Me^{(\omega + M\|R\|)t}.$$

(2) *Let $R_1, \ldots, R_n \in B(X)$. For $t \geq 0$ the map*
$$\mathbb{C}^n \to B(X), \ (z_1, \ldots, z_n) \mapsto e^{t(Z + z_1 R_1 + \cdots + z_n R_n)}$$
is analytical and for $i = 0, \ldots n$
$$\frac{d}{dz_i} e^{t(Z + z_1 R_1 + \cdots + z_n R_n)} = \int_0^t e^{(t-s)(Z + z_1 R_1 + \cdots + z_n R_n)} R_i e^{s(Z + z_1 R_1 + \cdots + z_n R_n)} \, ds.$$

PROOF. (1) follows from [**Da**], Th. 3.1 and the proof of it.
(2) The analyticity follows from (1).
For the formula it is enough to consider $n = 1$. Let $R := R_1$.
For $z_0 \in \mathbb{C}$ by (1)
$$e^{(Z+zR)t} = \sum_{n=0}^{\infty} (z-z_0)^n t^n \int_{\Delta^n} e^{u_0 t(Z+z_0 R)} R e^{u_1 t(Z+z_0 R)} R \ldots e^{u_n t(Z+z_0 R)} \, du_0 \ldots du_n.$$
This implies that
$$\frac{d}{dz} e^{(Z+zR)t}\big|_{z_0} = t \int_0^1 e^{u_0 t(Z+z_0 R)} R e^{(1-u_0) t(Z+z_0 R)} \, du_0.$$
□

The following proposition is known in the literature as Duhamel's principle:

PROPOSITION 5.4.5. *Let Z be the generator of a strongly continuous semigroup on X. Let $u \in C^1([0,\infty), X)$ such that $u(t) \in \operatorname{dom} Z$ and $\frac{d}{dt} u(t) - Zu(t) \in \operatorname{dom} Z$ for all $t \in [0, \infty)$. Then*
$$e^{tZ} u(0) - u(t) = -\int_0^t e^{sZ} \left(\frac{d}{dt} - Z\right) u(t-s) \, ds.$$

PROOF. see [**Ta**], Appendix A, (9.37) and (9.38). □

5.4.2. Square roots of generators and perturbations.

Assume that D is a densely defined closed operator on a Banach space X with bounded inverse and such that $-D^2$ is δ-sectorial.

There are well-defined fractional powers $(D^2)^\alpha$ for $\alpha \in \mathbb{R}$ ([**RR**], §11.4.2). These are densely defined closed operators that coincide for $\alpha \in \mathbb{Z}$ with the usual powers and satisfy
$$(D^2)^{\alpha+\beta} f = (D^2)^\alpha (D^2)^\beta f$$
for all $\alpha, \beta \in \mathbb{R}$ and $f \in \mathrm{dom}(D^2)^\gamma$ with $\gamma = \max\{\alpha, \beta, \alpha+\beta\}$. For $\alpha \leq 0$ the operator $(D^2)^\alpha$ is bounded and depends of α in a strongly continuous way.

In particular it follows that for $\alpha \geq 0$ the operator $(D^2)^{-\alpha}$ is a bounded inverse of $(D^2)^\alpha$.

Define $|D| := (D^2)^{\frac{1}{2}}$.

By [**Kat**], Th. 2, the operator $-|D|$ is $(\delta + \frac{\pi/2 - \delta}{2})$-sectorial and can be expressed in terms of the resolvents of D^2.

Note that for every $n \in \mathbb{N}$ the domain of $|D|^n$ is a core of $|D|$ and $\mathrm{dom}\, D^n$ is a core of D.

LEMMA 5.4.6. (1) *Let P be a closed densely defined operator on X and assume that P^2 is densely defined and $P^2|_M = P|_M$ for some dense subset M of $\mathrm{dom}\, P^2$.*

Then P is a bounded projection.

(2) *Let I be a closed densely defined operator on X and assume that I^2 is densely defined and that there is a dense subset M of $\mathrm{dom}\, I^2$ with $I^2|_M = 1|_M$.*

Then I is a bounded involution.

PROOF. (1) If $f \in M$, then $f = (1-P)f + Pf$ and $(1-P)Pf = P(1-P)f = 0$. We conclude that $M \subset \mathrm{Ker}\, P + \mathrm{Ker}(1-P) \subset \mathrm{dom}\, P$. If P is closed, then $\mathrm{Ker}\, P + \mathrm{Ker}(1-P)$ is closed, hence $\mathrm{Ker}\, P + \mathrm{Ker}(1-P) = X$, thus $\mathrm{dom}\, P = X$.

(2) The operator $P := \frac{1}{2}(1+I)$ is a closed projection on X in the sense of (1). Thus P is bounded. It follows that I is bounded as well. □

PROPOSITION 5.4.7. *The closure of the operator $|D|^{-1}D : \mathrm{dom}\, D \to X$ is a bounded involution I on X and*

(1) $\mathrm{dom}\, D = \mathrm{dom}\, |D|$ *and* $I(\mathrm{dom}\, D) \subset \mathrm{dom}\, D$.
(2) $|D| = ID = DI$ *and* $D = I|D| = |D|I$.

PROOF. The operator D^{-1} commutes with the resolvents of D^2. It follows that $|D|^{-1}D^{-1} = D^{-1}|D|^{-1}$. Hence $|D|^{-1}(\mathrm{dom}\, D) \subset \mathrm{dom}\, D$ because of $\mathrm{dom}\, D = D^{-1}X$, so $\mathrm{dom}\, D^2 \subset \mathrm{dom}(|D|^{-1}D)^2$.

If $f \in \mathrm{dom}\, D$, then $|D|^{-1}Df = D|D|^{-1}f$. For $f \in \mathrm{dom}\, D^2$ it follows that $(|D|^{-1}D)^2 f = f$.

Let $(f_n)_{n \in \mathbb{N}}$ be a sequence in $\mathrm{dom}\, D$ converging to zero. If $|D|^{-1}Df_n = D|D|^{-1}f_n$ converges, the limit is zero since D is closed. Hence $|D|^{-1}D$ is closable. By the previous lemma it extends to a bounded involution.

Since for $f \in \mathrm{dom}\, D^2$ we have that $DIf = |D|f$ and since DI is closed, the composition $IDI : \mathrm{dom}\, |D| \to X$ is well-defined and closed. It coincides with D on $\mathrm{dom}\, D^2$, hence it is a closed extension of D. It follows that $\mathrm{dom}\, D \subset \mathrm{dom}\, |D|$. The inclusion $\mathrm{dom}\, |D| \subset \mathrm{dom}\, D$ is shown analogously.

The equations are clear on $\mathrm{dom}\, D^2$, which is a core for D and $|D|$. □

The operator $P := \frac{1}{2}(1 + I)$ with I as in the previous proposition is a bounded projection on X. By the proposition $P \operatorname{dom} D \subset \operatorname{dom} D$ and P commutes with D and $|D|$.

From $ID = |D|$ and $I|D| = D$ it follows that $PD = -P|D|$. Thus with respect to the decomposition $X = PX \oplus (1-P)X$

$$D = \begin{pmatrix} PDP & 0 \\ 0 & -(1-P)D(1-P) \end{pmatrix}$$

and

$$|D| = \begin{pmatrix} PDP & 0 \\ 0 & (1-P)D(1-P) \end{pmatrix}.$$

By taking into account that $-|D|$ is $(\delta + \frac{\pi/2 - \delta}{2})$-sectorial it follows for the resolvent set of D:

PROPOSITION 5.4.8. *For $\lambda \in \mathbb{C}$*

$$\{\lambda, -\lambda\} \subset \rho(D) \Leftrightarrow \{\lambda, -\lambda\} \subset \rho(|D|).$$

Thus if $\lambda \in \mathbb{C}$ with $-\lambda^2 \in \Sigma_{\pi/2+\delta}$, then $\lambda \in \rho(D)$.

Furthermore for every $\delta' < \delta$ there is $C > 0$ such that for all λ with $-\lambda^2 \in \Sigma_{\delta' + \pi/2}$

$$\|(D-\lambda)^{-1}\| \leq \frac{C}{|\lambda|}.$$

COROLLARY 5.4.9. *Let $\omega \geq 0$ be such that there is $C > 0$ with $\|e^{-tD^2}\| \leq Ce^{-\omega t}$ for all $t \geq 0$.*

(1) *For every $\alpha \in \mathbb{R}$ and $\omega' < \omega$ there is $C > 0$ such that for all $t > 0$*

$$\||D|^\alpha e^{-tD^2}\| \leq C t^{-\alpha/2} e^{-\omega' t}.$$

(2) *For every $n \in \mathbb{N}$ and $\omega' < \omega$ there is $C > 0$ such that for all $t > 0$*

$$\|D^n e^{-tD^2}\| \leq C t^{-n/2} e^{-\omega' t}.$$

PROOF. The first assertion is [**RR**], Lemma 11.36, and the second one follows from the first one by $D = I|D|$ and $DI = ID$. □

PROPOSITION 5.4.10. *Let A be a bounded operator and let $\delta' < \delta$.*
  (1) *There is $R > 0$ such that $D + A - \lambda$ has a bounded inverse if $|\lambda| > R$ and $-\lambda^2 \in \Sigma_{\delta' + \pi/2}$.*
  (2) *There is $\omega > 0$ such that $-(D+A)^2 + \omega$ is δ'-sectorial.*
  (3) *$D + A$ commutes with $e^{-t(D+A)^2}$.*

PROOF. By Prop. 5.4.8 there is $M > 0$ such that for all λ with $-\lambda^2 \in \Sigma_{\delta' + \pi/2}$

$$\|(D-\lambda)^{-1}\| \leq \frac{M}{|\lambda|}.$$

Hence the Neumann series

$$(D + A - \lambda)^{-1} = (D - \lambda)^{-1} \sum_{n=0}^{\infty} (A(D-\lambda)^{-1})^n$$

converges for $|\lambda| > M\|A\|$ and $-\lambda^2 \in \Sigma_{\delta' + \pi/2}$.

This shows (1).

If $|\lambda| > 2M\|A\|$ and $-\lambda^2 \in \Sigma_{\delta'+\pi/2}$, then

$$\begin{aligned}
\|(D+A-\lambda)^{-1}\| &= \|\sum_{n=0}^{\infty}(D-\lambda)^{-1}\left(A(D-\lambda)^{-1}\right)^n\| \\
&\leq \sum_{n=0}^{\infty}\|A\|^n\|(D-\lambda)^{-1}\|^{n+1} \\
&\leq \sum_{n=0}^{\infty}\frac{M^{n+1}\|A\|^n}{|\lambda|^{n+1}} \\
&= \frac{M}{|\lambda|}(1-\frac{M\|A\|}{|\lambda|})^{-1} \\
&\leq \frac{2M}{|\lambda|}.
\end{aligned}$$

Let $\mu \in \{|\mu| > 4M^2\|A\|^2\} \cap \Sigma_{\delta'+\pi/2}$. If $\lambda \in \mathbb{C}$ with $-\lambda^2 = \mu$, then $\lambda \in \rho(D+A)$, hence the resolvent

$$(-(D+A)^2 - \mu)^{-1} = -(D+A-\lambda)^{-1}(D+A+\lambda)^{-1}$$

exists and is bounded by

$$\|(-(D+A)^2 - \mu)^{-1}\| \leq \frac{4M^2}{|\mu|}.$$

There is $\omega > 4M^2\|A\|^2$ such that

$$\Sigma_{\delta'+\pi/2} \cup \{0\} \subset \left(\{|\mu| > 4M^2\|A\|^2\} \cap \Sigma_{\delta'+\pi/2}\right) - \omega$$

and thus

$$\Sigma_{\delta'+\pi/2} \cup \{0\} \subset \rho\bigl(-(D+A)^2 + \omega\bigr).$$

Assertion (2) follows now from Lemma 5.4.1.

(3) follows from the fact that $e^{-t(D+A)^2}$ can be expressed in terms of the resolvents of $(D+A)^2$, which commute with $D+A$. $\square$

Bibliography

[BGV] N. Berline & E. Getzler & M. Vergne, *Heat Kernels and Dirac Operators* (Grundlehren der mathematischen Wissenschaften 298), Springer, 1996

[Bl] B. Blackadar, *K-Theory for Operator Algebras* (MSRI Publications 5), Springer, 1986

[Bo] J.-B. Bost, "Principe d'Oka, K-théorie et systèmes dynamiques non commutatifs", *Inv. Math.* 101 (1990), pp. 261–333

[Bu] U. Bunke, "On the gluing problem for the η-invariant", *J. Diff. Geom.* 41 (1995), no. 2, pp. 397–448.

[BK] U. Bunke & H. Koch, "The η-form and a generalized Maslov index", *manuscripta math.* 95 (1998), pp. 189–212

[CGT] J. Cheeger & M. Gromov & M. Taylor, "Finite Propagation Speed, Kernel Estimates for Functions of the Laplace Operator and the Geometry of Complete Riemannian Manifolds" *J. Diff. Geom.* 17 (1982), pp. 15–53

[CLM] S.E. Cappell & R. Lee & E.Y. Miller, "On the Maslov index", *Comm. Pure Appl. Math.* 47 (1994), no. 2, pp. 121–186

[CM] A. Connes & H. Moscovici, "Cyclic cohomology, the Novikov conjecture and hyperbolic groups", *Topology* 29 (1990), no. 3, pp. 345–388

[Co] J.B. Conway, *A Course in Functional Analysis* (Graduate Texts in Mathematics 96), Springer-Verlag, 1985

[CQ] J. Cuntz & D. Quillen, "Cyclic Homology and Nonsingularity", *J. Amer. Math. Soc.* 8 (1995), no. 2, pp. 373–442

[Da] E.B. Davies, *One-Parameter Semigroups*, (LMS Monographs 15), Academic Press, 1980

[Ka] M. Karoubi, "Homologie cyclique et K-théorie", *Astérisque* 149 (1987)

[Kat] T. Kato, "Note on Fractional Powers of Linear Operators", *Proc. Japan Academy* 36 (1960), pp. 94–97

[Kö] G. Köthe, *Topological Vector Spaces II* (Grundlehren der mathematischen Wissenschaften 237), Springer-Verlag, 1979

[La] E.C. Lance, *Hilbert C^*-modules*, (LMS Lecture Note Series 210), Cambridge University Press, 1995

[LLP] E. Leichtnam & J. Lott & P. Piazza, "On the homotopy invariance of higher signatures for manifolds with boundary", *J. Diff. Geom.* 54 (2000), no. 3, pp. 561–633

[LLK] E. Leichtnam & W. Lück & M. Kreck, "On the cut-and-paste property of higher signatures of a closed oriented manifold", *Topology* 41 (2002), no. 4, pp. 725–744

[LP1] E. Leichtnam & P. Piazza, "The b-pseudodifferential calculus on Galois coverings and a higher Atiyah-Patodi-Singer index theorem", *Mém. Soc. Math. Fr.* 68 (1997)

[LP2] E. Leichtnam & P. Piazza, "Spectral sections and higher Atiyah-Patodi-Singer index theory on Galois coverings. *Geom. Funct. Anal.* 8 (1998)", no. 1, pp. 17–58

[LP3] E. Leichtnam & P. Piazza, "Homotopy invariance of twisted higher signatures on manifolds with boundary", *Bull. Soc. Math. France* 127 (1999), no. 2, pp. 307–331

[LP4] E. Leichtnam & P. Piazza, "A higher Atiyah-Patodi-Singer index theorem for the signature operator on Galois coverings", *Ann. Global Anal. Geom.* 18 (2000), no. 2, pp. 171–189

[LV] G. Lion & M. Vergne, *The Weil Representation, Maslov index and theta series* (Progress in Mathematics 6), Birkhäuser, 1980

[Lo1] J. Lott, "Higher eta-invariants", *K-Theory* 6 (1992), no.3, pp. 191–233

[Lo2] J. Lott, "Superconnections and higher index theory", *Geom. Funct. Anal.* 2 (1992), no. 4, pp. 421–454

[Lo3] J. Lott, "Diffeomorphisms and Noncommutative Analytic Torsion", *Mem. Am. Math. Soc.* 673 (1999)

[Ma] A. Mallios, *Topological Algebras* (Mathematics Studies 124), North-Holland, 1986
[Me] R.B. Melrose, *The Atiyah-Patodi-Singer index theorem* (Research Notes in Mathematics 4), A K Peters, 1993.
[MF] A.S. Miščenko & A.T. Fomenko, "The Index of Elliptic Operators over C^*-Algebras", *Math. USSR Izvestija* 15 (1980), no. 1, pp. 87–112
[PS] P. Piazza & T. Schick, "Bordism, rho-invariants and the Baum-Connes conjecture" (2005), preprint, http://arXiv.org/math.KT/0407388
[RR] M. Renardy & R.C. Rogers, *An Introduction to Partial Differential Equations* (Texts in Applied Mathematics 13), Springer-Verlag, 1993
[Ro] J. Roe, *Elliptic Operators, Topology and Asymptotic Methods* (Pitman Research Notes in Mathematics Series 395), Longman, 1998
[Ros1] J. Rosenberg, "C^*-algebras, positive scalar curvature, and the Novikov conjecture" *Inst. Hautes Etudes Sci. Publ. Math.* 58 (1983), pp. 197–212
[Ros2] J. Rosenberg, "Analytic Novikov for Topologists", *Novikov Conjectures, Index Theorems and Rigidity (1)* (LMS Lecture Note Series 226), Cambridge University Press, 1995, pp. 338–372
[Se] Z. Semadeni, *Schauder Bases in Banach Spaces of Continuous Functions* (Lecture Notes in Mathematics 918), Springer-Verlag, 1982
[Ta] M.E. Taylor, *Partial Differential Equations I* (Applied Mathematical Sciences 115), Springer-Verlag, 1996
[Tr] F. Treves, *Topological Vector Spaces, Distributions and Kernels* (Pure and Applied Mathematics 25), Academic Press, 1967
[Wa] C.T.C. Wall, "Non-Additivity of the Signature", *Inv. Math.* 7 (1969), pp. 269–274
[WO] N.E. Wegge-Olson, *K-Theory and C^*-Algebras*, Oxford University Press, 1993
[Wu] F. Wu, "A bivariant Chern character and the higher Γ-index theorem", *K-Theory* 11 (1997), no. 2, pp. 35–82

Editorial Information

To be published in the *Memoirs*, a paper must be correct, new, nontrivial, and significant. Further, it must be well written and of interest to a substantial number of mathematicians. Piecemeal results, such as an inconclusive step toward an unproved major theorem or a minor variation on a known result, are in general not acceptable for publication.

Papers appearing in *Memoirs* are generally at least 80 and not more than 200 published pages in length. Papers less than 80 or more than 200 published pages require the approval of the Managing Editor of the Transactions/Memoirs Editorial Board.

As of May 31, 2007, the backlog for this journal was approximately 15 volumes. This estimate is the result of dividing the number of manuscripts for this journal in the Providence office that have not yet gone to the printer on the above date by the average number of monographs per volume over the previous twelve months, reduced by the number of volumes published in four months (the time necessary for preparing a volume for the printer). (There are 6 volumes per year, each usually containing at least 4 numbers.)

A Consent to Publish and Copyright Agreement is required before a paper will be published in the *Memoirs*. After a paper is accepted for publication, the Providence office will send a Consent to Publish and Copyright Agreement to all authors of the paper. By submitting a paper to the *Memoirs*, authors certify that the results have not been submitted to nor are they under consideration for publication by another journal, conference proceedings, or similar publication.

Information for Authors

Memoirs are printed from camera copy fully prepared by the author. This means that the finished book will look exactly like the copy submitted.

Initial submission. The AMS uses Centralized Manuscript Processing for initial submissions. Authors should submit a PDF file using the Initial Manuscript Submission form found at www.ams.org/cgi-bin/peertrack/submission.pl, or send one copy of the manuscript to the following address: Centralized Manuscript Processing, MEMOIRS OF THE AMS, 201 Charles Street, Providence, RI 02904-2294 USA. If a paper copy is being forwarded to the AMS, indicate that it is for it Memoirs and include the name of the corresponding author, contact information such as email address or mailing address, and the name of an appropriate Editor to review the paper (see the list of Editors below).

The paper must contain a *descriptive title* and an *abstract* that summarizes the article in language suitable for workers in the general field (algebra, analysis, etc.). The *descriptive title* should be short, but informative; useless or vague phrases such as "some remarks about" or "concerning" should be avoided. The *abstract* should be at least one complete sentence, and at most 300 words. Included with the footnotes to the paper should be the 2000 *Mathematics Subject Classification* representing the primary and secondary subjects of the article. The classifications are accessible from www.ams.org/msc/. The list of classifications is also available in print starting with the 1999 annual index of *Mathematical Reviews*. The Mathematics Subject Classification footnote may be followed by a list of *key words and phrases* describing the subject matter of the article and taken from it. Journal abbreviations used in bibliographies are listed in the latest *Mathematical Reviews* annual index. The series abbreviations are also accessible from www.ams.org/publications/. To help in preparing and verifying references, the AMS offers MR Lookup, a Reference Tool for Linking, at www.ams.org/mrlookup/.

Electronically prepared manuscripts. The AMS encourages electronically prepared manuscripts, with a strong preference for $\mathcal{A}_{\mathcal{M}}\mathcal{S}$-LaTeX. To this end, the Society has prepared $\mathcal{A}_{\mathcal{M}}\mathcal{S}$-LaTeX author packages for each AMS publication. Author packages include instructions for preparing electronic manuscripts, samples, and a style file that generates the particular design specifications of that publication series. Though $\mathcal{AMS}$-LaTeX is the highly preferred format of TeX, author packages are also available in $\mathcal{AMS}$-TeX.

Authors may retrieve an author package from the AMS website starting from `www.ams.org/tex/` or via FTP to `ftp.ams.org` (login as `anonymous`, enter username as password, and type `cd pub/author-info`). The *AMS Author Handbook* and the *Instruction Manual* are available in PDF format following the author packages link from `www.ams.org/tex/`. The author package can also be obtained free of charge by sending email to `tech-support@ams.org` (Internet) or from the Publication Division, American Mathematical Society, 201 Charles St., Providence, RI 02904-2294, USA. When requesting an author package, please specify $\mathcal{AMS}$-LaTeX or $\mathcal{AMS}$-TeX and the publication in which your paper will appear. Please be sure to include your complete mailing address.

After acceptance. The final version of the electronic file should be sent to the Providence office (this includes any TeX source file, any graphics files, and the DVI or PostScript file) immediately after the paper has been accepted for publication.

Before sending the source file, be sure you have proofread your paper carefully. The files you send must be the EXACT files used to generate the proof copy that was accepted for publication. For all publications, authors are required to send a printed copy of their paper, which exactly matches the copy approved for publication, along with any graphics that will appear in the paper.

Accepted electronically prepared files can be submitted via the web at `www.ams.org/submit-book-journal/`, sent via FTP, or sent on CD-Rom or diskette to the Electronic Prepress Department, American Mathematical Society, 201 Charles Street, Providence, RI 02904-2294 USA. TeX source files, DVI files, and PostScript files can be transferred over the Internet by FTP to the Internet node `ftp.ams.org` (130.44.1.100). When sending a manuscript electronically via CD-Rom or diskette, please be sure to include a message identifying the paper as a Memoir.

Electronically prepared manuscripts can also be sent via email to `pub-submit@ams.org` (Internet). In order to send files via email, they must be encoded properly. (DVI files are binary and PostScript files tend to be very large.)

Electronic graphics. Comprehensive instructions on preparing graphics are available at `www.ams.org/jourhtml/`. A few of the major requirements are given here.

Submit files for graphics as EPS (Encapsulated PostScript) files. This includes graphics originated via a graphics application as well as scanned photographs or other computer-generated images. If this is not possible, TIFF files are acceptable as long as they can be opened in Adobe Photoshop or Illustrator. No matter what method was used to produce the graphic, it is necessary to provide a paper copy to the AMS.

Authors using graphics packages for the creation of electronic art should also avoid the use of any lines thinner than 0.5 points in width. Many graphics packages allow the user to specify a "hairline" for a very thin line. Hairlines often look acceptable when proofed on a typical laser printer. However, when produced on a high-resolution laser imagesetter, hairlines become nearly invisible and will be lost entirely in the final printing process.

Screens should be set to values between 15% and 85%. Screens which fall outside of this range are too light or too dark to print correctly. Variations of screens within a graphic should be no less than 10%.

Inquiries. Any inquiries concerning a paper that has been accepted for publication should be sent to `memo-query@ams.org` or directly to the Electronic Prepress Department, American Mathematical Society, 201 Charles St., Providence, RI 02904-2294 USA.

Editors

This journal is designed particularly for long research papers, normally at least 80 pages in length, and groups of cognate papers in pure and applied mathematics. Papers intended for publication in the *Memoirs* should be addressed to one of the following editors. The AMS uses Centralized Manuscript Processing for initial submissions to AMS journals. Authors should follow instructions listed on the Initial Submission page found at www.ams.org/memo/memosubmit.html.

Algebra to ALEXANDER KLESHCHEV, Department of Mathematics, University of Oregon, Eugene, OR 97403-1222; email: ams@noether.uoregon.edu

Algebraic geometry and its application to MINA TEICHER, Emmy Noether Research Institute for Mathematics, Bar-Ilan University, Ramat-Gan 52900, Israel; email: teicher@macs.biu.ac.il

Algebraic geometry to DAN ABRAMOVICH, Department of Mathematics, Brown University, Box 1917, Providence, RI 02912; email: amsedit@math.brown.edu

Algebraic number theory to V. KUMAR MURTY, Department of Mathematics, University of Toronto, 100 St. George Street, Toronto, ON M5S 1A1, Canada; email: murty@math.toronto.edu

Algebraic topology to ALEJANDRO ADEM, Department of Mathematics, University of British Columbia, Room 121, 1984 Mathematics Road, Vancouver, British Columbia, Canada V6T 1Z2; email: adem@math.ubc.ca

Combinatorics to JOHN R. STEMBRIDGE, Department of Mathematics, University of Michigan, Ann Arbor, Michigan 48109-1109; email: FRS@umich.edu

Complex analysis and harmonic analysis to ALEXANDER NAGEL, Department of Mathematics, University of Wisconsin, 480 Lincoln Drive, Madison, WI 53706-1313; email: nagel@math.wisc.edu

Differential geometry and global analysis to LISA C. JEFFREY, Department of Mathematics, University of Toronto, 100 St. George St., Toronto, ON Canada M5S 3G3; email: jeffrey@math.toronto.edu

Dynamical systems and ergodic theory to AMIE WILKINSON, Department of Mathematics, Northwestern University, 2033 Sheridan Road, Evanston, IL 60208-2730; email: transactions@math.northwestern.edu

Functional analysis and operator algebras to DIMITRI SHLYAKHTENKO, Department of Mathematics, University of California, Los Angeles, CA 90095; email: shlyakht@math.ucla.edu

Geometric analysis to WILLIAM P. MINICOZZI II, Department of Mathematics, Johns Hopkins University, 3400 N. Charles St., Baltimore, MD 21218; email: trans@math.jhu.edu

Geometric analysis to MLADEN BESTVINA, Department of Mathematics, University of Utah, 155 South 1400 East, JWB 233, Salt Lake City, Utah 84112-0090; email: bestvina@math.utah.edu

Harmonic analysis, representation theory, and Lie theory to ROBERT J. STANTON, Department of Mathematics, The Ohio State University, 231 West 18th Avenue, Columbus, OH 43210-1174; email: stanton@math.ohio-state.edu

Logic to STEFFEN LEMPP, Department of Mathematics, University of Wisconsin, 480 Lincoln Drive, Madison, Wisconsin 53706-1388; email: lempp@math.wisc.edu

Partial differential equations to GUSTAVO PONCE, Department of Mathematics, South Hall, Room 6607, University of California, Santa Barbara, CA 93106; email: ponce@math.ucsb.edu

Partial differential equations and dynamical systems to PETER POLACIK, School of Mathematics, University of Minnesota, Minneapolis, MN 55455; email: polacik@math.umn.edu

Probability and statistics to KRZYSZTOF BURDZY, Department of Mathematics, University of Washington, Box 354350, Seattle, Washington 98195-4350; email: burdzy@math.washington.edu

Real analysis and partial differential equations to DANIEL TATARU, Department of Mathematics, University of California, Berkeley, Berkeley, CA 94720; email: tataru@math.berkeley.edu

All other communications to the editors should be addressed to the Managing Editor, ROBERT GURALNICK, Department of Mathematics, University of Southern California, Los Angeles, CA 90089-1113; email: guralnic@math.usc.edu.

Titles in This Series

887 **Charlotte Wahl,** Noncommutative Maslov index and eta-forms, 2007
886 **Robert M. Guralnick and John Shareshian,** Symmetric and alternating groups as monodromy groups of Riemann surfaces I: Generic covers and covers with many branch points, 2007
885 **Jae Choon Cha,** The structure of the rational concordance group of knots, 2007
884 **Dan Haran, Moshe Jarden, and Florian Pop,** Projective group structures as absolute Galois structures with block approximation, 2007
883 **Apostolos Beligiannis and Idun Reiten,** Homological and homotopical aspects of torsion theories, 2007
882 **Lars Inge Hedberg and Yuri Netrusov,** An axiomatic approach to function spaces, spectral synthesis and Luzin approximation, 2007
881 **Tao Mei,** Operator valued Hardy spaces, 2007
880 **Bruce C. Berndt, Geumlan Choi, Youn-Seo Choi, Heekyoung Hahn, Boon Pin Yeap, Ae Ja Yee, Hamza Yesilyurt, and Jinhee Yi,** Ramanujan's forty identities for Rogers-Ramanujan functions, 2007
879 **O. García-Prada, P. B. Gothen, and V. Muñoz,** Betti numbers of the moduli space of rank 3 parabolic Higgs bundles, 2007
878 **Alessandra Celletti and Luigi Chierchia,** KAM stability and celestial mechanics, 2007
877 **María J. Carro, José A. Raposo, and Javier Soria,** Recent developments in the theory of Lorentz spaces and weighted inequalities, 2007
876 **Gabriel Debs and Jean Saint Raymond,** Borel liftings of Borel sets: Some decidable and undecidable statements, 2007
875 **C. Krattenthaler and T. Rivoal,** Hypergéométrie et fonction zêta de Riemann, 2007
874 **Sonia Natale,** Semisolvability of semisimple Hopf algebras of low dimension, 2007
873 **A. J. Duncan,** Exponential genus problems in one-relator products of groups, 2007
872 **Anthony V. Geramita, Tadahito Harima, Juan C. Migliore, and Yong Su Shin,** The Hilbert function of a level algebra, 2007
871 **Pascal Auscher,** On necessary and sufficient conditions for L^p-estimates of Riesz transforms associated to elliptic operators on $\mathbb{R}^n$ and related estimates, 2007
870 **Takuro Mochizuki,** Asymptotic behaviour of tame harmonic bundles and an application to pure twistor D-modules, Part 2, 2007
869 **Takuro Mochizuki,** Asymptotic behaviour of tame harmonic bundles and an application to pure twistor D-modules, Part 1, 2007
868 **Gelu Popescu,** Entropy and multivariable interpolation, 2006
867 **Vilmos Totik,** Metric properties of harmonic measures, 2006
866 **William Craig,** Semigroups underlying first-order logic, 2006
865 **Nathanial P. Brown,** Invariant means and finite representation theory of $C*$-algebras, 2006
864 **John M. Lee,** Fredholm operators and Einstein metrics on conformally compact manifolds, 2006
863 **M. Lübke and A. Teleman,** The Universal Kobayashi-Hitchin correspondence on Hermitian manifolds, 2006
862 **Alberto Canonaco,** The Beilinson complex and canonical rings of irregular surfaces, 2006
861 **Leon A. Takhtajan and Lee-Peng Teo,** Weil-Petersson metric on the universal Teichmüller space, 2006

For a complete list of titles in this series, visit the
AMS Bookstore at **www.ams.org/bookstore/**.